Security Risk Models for Cyber Insurance

Security Risk Models for Cyber Insurance

Edited by
David Rios Insua
Caroline Baylon
Jose Vila

CRC Press
Taylor & Francis Group
Boca Raton London New York

CRC Press is an imprint of the
Taylor & Francis Group, an **informa** business

A CHAPMAN & HALL BOOK

First edition published 2021
by CRC Press
6000 Broken Sound Parkway NW, Suite 300, Boca Raton, FL 33487-2742

and by CRC Press
2 Park Square, Milton Park, Abingdon, Oxon, OX14 4RN

© 2021 Taylor & Francis Group, LLC

CRC Press is an imprint of Taylor & Francis Group, LLC

The right of David Rios Insua, Caroline Baylon and Jose Vila to be identified as the authors of the editorial material, and of the authors for their individual chapters, has been asserted in accordance with sections 77 and 78 of the Copyright, Designs and Patents Act 1988.

ISBN: 9780367339494 (hbk)
ISBN: 9780429329487(ebk)

Typeset in Computer Modern font
by KnowledgeWorks Global Ltd.

To Susana, Isa, and Carla. DAVID RIOS INSUA
To my parents. CAROLINE BAYLON
To Julian, Oriana, Marina, and Rodrigo. JOSE VILA

Contents

Foreword

Cybersecurity has firmly established itself as a major global threat. We regularly hear reports of a company having experienced "the biggest data breach in history," with each much larger than the last. It is not uncommon for organisations to suffer attacks involving the data of hundreds of millions—or even billions—of customers. We are also seeing a rise in cyber attacks on critical infrastructure, from transportation networks to the power grid, and their related potential for disruption. As a response to this risk, the insurance industry is developing novel cyber insurance products that facilitate risk transfer within a risk management portfolio.

However, the development of cyber insurance products presents a number of challenges. One of them is the rapidly evolving cyber threat landscape, including the growth in the number of attacks and the sophistication of attackers, that makes it difficult to accurately assess cyber risks. Another is the limited amount of historical data, which is traditionally the basis for insurance underwriting. In addition, customers are often unable to fully evaluate the cybersecurity risks they face and lack clarity around cyber insurance options. These issues call for new approaches in this domain.

This publication is the result of an initiative of the AXA-ICMAT Chair in Adversarial Risk Analysis, supported by the AXA Research Fund, and the CYBECO (Supporting Cyber Insurance from a Behavioural Choice Perspective) Project, a European Union-funded project through the Horizons 2020 programme. The project has brought together a diverse team of interdisciplinary European researchers, including cybersecurity practitioners as well as experts from the fields of risk analysis, psychology, behavioural economics, decision analysis, computer science, modelling, and policymaking. The findings underscore the importance of supporting independent academic research projects to find new ways to tackle the cybersecurity challenges that our society faces today.

The result is a volume that examines cyber insurance decision-making processes within organisations, identifies the behavioural issues underlying cybersecurity, and proposes innovative risk analysis models. It provides a timely contribution to the literature on cybersecurity and cyber insurance, offering guidance to companies in their cybersecurity resource allocation decisions and insights for insurers and brokers in their risk mitigation roles, thus contributing to a more resilient and "cybersecure" environment.

MARIE BOGATAJ, Director of the AXA Research Fund, Paris
ARNAUD TANGUY, AXA Group Chief Security Officer, Paris

Preface

A defining feature of modern society is its pervasive digitalisation, as exemplified by the information systems that store and process valuable data, much of it confidential. These systems include cyber-physical systems that operate critical infrastructure, the social networks that host so many of our interactions with others, and platforms that enable financial transactions such as online shopping or banking, to name but a few. Against this backdrop, cyber attacks are increasing in frequency, impact, and sophistication and can affect all types of organisations from corporations and governments to SMEs and NGOs, as well as individual citizens. The number of security breaches has increased by 67% in the past five years and cybercrime is estimated to cost the world economy $600 billion annually, or 0.8% of global GDP (Accenture and Ponemon Institute, 2019; McAfee and Center for Strategic and International Studies, 2018). The 2017 NotPetya attack was particularly destructive, causing over $10 billion in damage as it propagated across the corporate networks of a number of major multinational companies.

The use of cybersecurity risk management methods is essential in order to deal with these challenges. These methods enable organisations to assess the threats to their assets, what security measures they should implement to reduce the likelihood of such threats occurring, and to lessen their potential impacts should they occur. Yet, despite their virtues, the current frameworks used for cybersecurity risk management are mainly based on risk matrices, which have well-documented shortcomings that could potentially lead to a suboptimal allocation of cybersecurity resources. They also do not typically take into account the intentionality of threats. This may be even more of an issue if we take into account the increasing variety of threats, as well as the growing number of security measures to choose from to counter these threats.

In this context, new methods of cybersecurity risk management are emerging, notably involving the use of cyber insurance. Cyber insurance can fulfil a key role by keeping risks manageable for insured companies by transferring the risk to insurers. It also provides companies with incentives to improve their cybersecurity by requiring them to implement certain minimum protections, thereby reducing overall risk. However, the cyber insurance market is still underdeveloped for several reasons. From the demand side, companies may struggle to decide whether or not to buy cyber insurance and which products to buy, in part due to difficulty understanding their cybersecurity risk. This is made more complex by the rapidly changing nature of the risk. From the supply side, this also means that it is difficult for insurance companies to create an overall risk picture for the domain, making it challenging to design and price cyber insurance products.

To this end, this book presents findings from the European Union-funded CYBECO (Supporting Cyber Insurance from a Behavioural Choice Perspective) Project, which has developed new models for risk management in cybersecurity, including a model to assist companies in selecting cyber insurance products and models to aid insurers in underwriting them. More specifically, the models developed consider the behavioural choices of both companies (their risk reduction and risk transfer decisions) and insurers (their risk assessment decisions). They also take into account the behavioural choices of the relevant threat actors (their risk generation decisions). This book facilitates the adoption of these models as well, by providing a case study to illustrate how to implement the cyber insurance product selection model and a link to a prototype "toolbox" based on the model that assesses a company's cybersecurity risk and provides advice on their optimal allocation of financial resources between cybersecurity and cyber insurance. In this way, this book aims to promote better cybersecurity risk management practices and the greater uptake of cyber insurance, thus helping to reduce overall cybersecurity risk and benefitting society as a whole.

DAVID RIOS INSUA, Valdoviño
CAROLINE BAYLON, London
JOSE VILA, Valencia

Acknowledgements

We would like to express our gratitude to the European Commission for its generous funding of the CYBECO (Supporting Cyber Insurance from a Behavioural Choice Perspective) Project through its Horizons 2020 programme under grant agreement number 740920.

In addition, we would like to thank AXA for its valuable suggestions regarding this book. We are particularly grateful to Marie Bogataj, Head of the AXA Research Fund; Arnaud Tanguy, Group Chief Security Officer; Dr Cecile Wendling, Head of Strategy, Threat Anticipation, and Research; Scott Sayce, Group Head of Financial Lines and Cyber; Thomas Lawson, Information Risk Advisory Director; and Sara Gori, Global Head of Reputation Risk Management, whose comments on drafts of this book were invaluable. Any remaining errors are our own.

Prof. David Rios Insua also appreciates the support of the AXA Research Fund through the AXA-ICMAT Chair in Adversarial Risk Analysis, of the US National Science Foundation through grant number DMS-163851 at SAMSI (The Statistical and Applied Mathematical Sciences Institute), as well as of the Spanish Ministry of Science through project number MTM2017-86875-C3-1-R.

The views set forth in this volume are those of the authors and should not be taken to represent the perspectives of the European Commission, of AXA, of any other organisations mentioned in the book.

DAVID RIOS INSUA, Valdoviño
CAROLINE BAYLON, London
JOSE VILA, Valencia

List of Figures

List of Tables

Editors

David Ríos Insua is AXA-ICMAT Chair in Adversarial Risk Analysis and a Member of the Spanish Royal Academy of Sciences.

Caroline Baylon is Security Research and Innovation Lead at AXA and a Research Affiliate at the Centre for the Study of Existential Risk, University of Cambridge.

José Vila is Scientific Director at DevStat, Associate Professor at the University of Valencia, and Research Fellow at the Centre for Research on Social and Economic Behaviour (ERI-CES) and Intelligent Data Analysis Laboratory (IDAL).

Contributors

Sebastain Awondo
The University of Alabama
Tuscaloosa, USA

Caroline Baylon
AXA
London, UK

Dawn Branley-Bell
Northumbria University
Newcastle upon Tyne, UK

Pam Briggs
Northumbria University
Newcastle upon Tyne, UK

Aitor Couce Vieira
ICMAT-CSIC
Madrid, Spain

Lynne Coventry
Northumbria University
Newcastle upon Tyne, UK

Michel van Eeten
Delft University of Technology
Delft, Netherlands

Yolanda Gomez
DevStat
Valencia, Spain

Katsiaryna Labunets
Delft University of Technology
Delft, Netherlands

Inés Martínez
Delft University of Technology
Delft, Netherlands

Wolter Pieters
Delft University of Technology
Delft, Netherlands

Alberto Redondo
ICMAT-CSIC
Madrid, Spain

David Ríos Insua
ICMAT-CSIC
Madrid, Spain

Jhoties Sewnandan
Delft University of Technology
Delft, Netherlands

Deepak Subramanian
AXA
Paris, France

Nikos Vasileiadis
TREK
Thessaloniki, Greece

Jose Vila
DevStat and University of Valencia
Valencia, Spain

Abbreviations

Abbreviations

ABM   Agent-Based Model
ARA   Adversarial Risk Analysis
BAID   Bi-Agent Influence Diagram
BEE   Behavioural Economics Experiment
CSRM   Cybersecurity Risk Management
CVaR   Conditional Value-at-Risk
DDoS   Distributed Denial-of-Service
DNS   Domain Name System
EIOPA   European Informational and Occupational Pensions Authority
ENISA   European Union Agency for Cybersecurity
EPPM   Extended Parallel Process Model
EU   European Union
GDP   Gross Domestic Product

GDPR   General Data Protection Regulation
ID   Influence Diagram
IoT   Internet of Things
ISF   Information Security Forum
IT   Information Technology
LDA   Loss Distribution Approach
MAID   Multi-Agent Influence Diagram
NGO   Non-Governmental Organisation
NIS   Network and Information Systems
PII   Personal Identifiable Information
PMT   Protection Motivation Theory
PR   Public Relations
SME   Small and Medium Enterprise
TAID   Tri-Agent Influence Diagram
TPB   Theory of Planned Behaviour
VaR   Value-at-Risk

1

Introduction

David Ríos Insua
ICMAT

Nikos Vasileiadis
TREK

Aitor Couce Vieira
ICMAT

Caroline Baylon
AXA

CONTENTS

This chapter begins by presenting the central thesis of this book: In order to tackle the pressing cybersecurity challenge, companies need to employ a reliable Cybersecurity Risk Management (CSRM) methodology. Yet current CSRM approaches have significant shortcomings that can lead to the incorrect prioritisation of cyber risks and of resources. The inclusion of cyber insurance could significantly improve CSRM methods, but the cyber insurance market is currently underdeveloped due to both demand and supply side challenges. To overcome this, this book proposes new models for risk management in cybersecurity, including a main CSRM model for companies and a series of auxiliary models for insurers. The following sections in this chapter introduce fundamental concepts that later parts of the book will build upon, first providing a schematic view of the factors involved in CSRM and then describing the key facets of the cyber insurance market at present.

1.1 Overview

1.1.1 The cyber threat landscape

The threat actors

Cybersecurity is a major global concern, with attacks becoming increasingly ubiquitous, growing in both frequency and size (WEF, 2020). There are a diversity of threat actors whose numbers are steadily rising as well. These include *hacktivists*, who are closely linked

1

with political or social movements and could involve anyone from hackers taking action to defend free speech to those closely aligned with terrorist organisations. *Insiders* are another important cyber threat and, indeed, the biggest source of incidents (Cardenas et al., 2009). However, they may be the easiest to handle through a sound cybersecurity program. *Cybercriminals* are increasing in capability. Many cybercriminal groups have become mature professional organisations, with some employing dozens of hackers and possessing extensive financial resources (Cardenas et al., 2008). Well-functioning markets on the "dark web" provide skilled individuals with incentives to steal data or develop new automated attack tools (Herley and Florêncio, 2010). The ability to purchase such tools has also made it easier for those without advanced technical skills to engage in cybercrime. Perhaps the most formidable threats at present are *nation states*. Although partially constrained by the possible military, economic, and political repercussions of launching cyber attacks, state actors are increasingly developing offensive programs and stockpiling cyberweapons, which could be released either accidentally or intentionally. This is a particular concern given the increased tensions between global powers at present.

A rise in the number and impact of attacks

As companies, governments, and individuals become ever more connected to the internet, the attack surface is growing and along with it the number and impact of attacks as well. High-profile corporate data breaches in recent years include the 2017 breach of Equifax, in which the data of over 140 million customers—including social security and credit card numbers—was stolen. The Yahoo data breach, first reported in 2016 but dating back to 2013, saw the theft of passwords as well as personal data associated with all 3 billion of its user accounts. The 2015 breach of Anthem resulted in the theft of 78.8 million client records containing Personally Identifiable Information (PII). In the 2013 Target data breach, hackers were able to access the Target network through an attack on one of its third party suppliers, an air conditioning company; they made off with the credit card information of 70 million customers and also caused Target major reputational damage (Manworren et al., 2016).

Among a spate of major ransomware attacks, the 2017 WannaCry attack took down the UK National Health Service, Telefonica, and FedEx, as well as others, causing significant disruption and entailing losses estimated to have reached $4 billion (Berr, 2016). Its use of a leaked US National Security Agency exploit made it particularly damaging. Governments have also been hard hit, with a 2018 ransomware attack on the City of Atlanta impacting city services, from utilities to parking, that took months to recover from. Similarly, a 2016 ransomware attack on San Francisco public transit disrupted payment services for the city's light rail system.

The 2017 NotPetya attack affected thousands of companies including Maersk, DHL, and Saint-Gobain and caused an estimated $10 billion in damages (Greenberg, 2018). Although purporting to be ransomware, many experts believe that NotPetya was in fact a cyberweapon created by Russia and targeted at Ukraine that inadvertently hit a number of unrelated targets. Other high-profile attacks attributed to state actors include the 2015 attack on the Ukrainian power grid that left some 230,000 people without power for up to six hours, an attack that Russia is also thought to have instigated. The first known successful attack on a power grid, it illustrates the rise of attacks on cyber-physical systems with real world consequences. An early example was the 2010 Stuxnet attack on an Iranian nuclear facility that damaged one fifth of its nuclear centrifuges, which is widely believed to have been carried out by the US and Israel (Brenner, 2013).

Additionally, distributed denial-of-service (DDoS) attacks are growing more destructive, in large part due to the exponential growth of the Internet of Things (IoT); many IoT devices are rolled out quickly, cheaply, with little thought as to cybersecurity, and therefore can be readily co-opted into botnets. The 2016 Mirai botnet, composed of a host of internet-connected devices from cameras to baby monitors, took down major internet sites including Twitter, Netflix, CNN, and The New York Times by launching an attack on Dyn, which controls much of the internet domain name system.

Finally, new types of attacks are regularly emerging. For example, the rise in value of cryptocurrencies has brought about a growth in cryptojacking attacks that take over computers to secretly mine bitcoin. And as progress is made in AI, cybercriminals are increasingly employing AI-enabled attacks as well.

1.1.2 Cybersecurity risk management

To deal with these challenges, the use of a sound Cybersecurity Risk Management (CSRM) methodology is essential. CSRM techniques rely heavily on risk analysis (Bedford and Cooke, 2001), enabling organisations to assess the risks to their assets as well as what safeguards should be implemented to reduce the likelihood of various threats occurring and their impact if they do. Numerous frameworks have been developed to support cybersecurity risk management, including the international standard ISO 27005 (ISO, 2011), CRAMM in the UK (Central Communication and Telecommunication Agency, 2003), MAGERIT in Spain (Amutio et al., 2012), EBIOS in France (ANSSI, 2010), the NIST Risk Management Framework and others in the US (NIST, 2018; NIST, 2012), and CORAS by an EU-funded project (Lund et al., 2011). Similarly, a number of compliance and control assessment frameworks like ISO 27001 (ISO, 2013), Common Criteria (Common Criteria, 2017), and the Cloud Controls Matrix (Cloud Security Alliance, 2019) offer guidance on the implementation of cybersecurity best practices. The above methodologies and frameworks provide an extensive catalogue of threats, assets, and controls, as well as detailed guidelines for the implementation of countermeasures to protect digital assets. However, much remains to be done regarding risk analysis from a methodological point of view.

Challenges with current risk management approaches in cybersecurity

A detailed study of the main approaches to CSRM reveals that they often rely on risk matrices, which have well-documented shortcomings (Cox, 2008; Thomas et al., 2013). Compared with more stringent methods, the ordinal ratings for likelihood, severity, and risk used in risk matrices are prone to ambiguity and subjective interpretation. They also systematically assign the same rating to threats that are significantly different qualitatively. This can potentially lead to a sub-optimal allocation of cybersecurity resources. Hubbard and Seiersen (2016) and Allodi and Massacci (2017) provide additional critical perspectives on the use of risk matrices in cybersecurity. The problem may be even more significant if we take into account the increasing variety of cybersecurity threats, as well as the growing complexity of the security controls used in cybersecurity risk management.

Moreover, these methodologies typically do not explicitly take into account the intentionality of certain threats, with a few exceptions like the UK's IS1 (National Technical Authority for Information Assurance, 2012). Yet the vast majority of security companies and industry bodies emphasise the importance of defending against adversarial threats, not just accidental or environmental ones (ISF, 2017). As a consequence, current CSRM approaches may lead companies to incorrectly prioritise cyber risks and the measures they should implement to defend against them.

Cyber insurance as part of an alternative CSRM methodology and obstacles to overcome

In this context, a complementary way of dealing with cyber risks through risk transfer is emerging. This involves the use of cyber insurance products, which have been introduced in recent years by companies like AXA, Generali, or Allianz. Cyber insurance can fulfil a key role in the economics of cybersecurity in several ways. First, by keeping cyber risks manageable for insured companies by transferring the risk to the insurance provider. Second, by providing incentives to improve cybersecurity, requiring companies to implement certain minimum protections, thereby reducing overall risk.

Unfortunately, cyber insurance is still underdeveloped for a variety of reasons. On the demand side, companies often struggle to decide whether or not to buy insurance, and which products to buy. On the supply side, it is difficult for insurance companies to assess the overall risk when it comes to cybersecurity and thus to design their product offerings, partly because of a lack of data. This is discussed further in Section 1.3.

1.1.3   The approach of this book

The growing cyber threat landscape, coupled with the shortcomings of current CSRM frameworks and the unrealised potential of cyber insurance for risk management, underscores the need for new cybersecurity risk management approaches. This book presents findings from the European Union-funded CYBECO (Supporting Cyber Insurance from a Behavioural Choice Perspective) Project, which has developed new models for risk management in cybersecurity. This includes both a model for companies and a series of models for insurers, in order to help further develop both the demand and supply sides of the cyber insurance market. More specifically, the model aimed at companies assists them in determining their optimal cybersecurity resource allocation (including selecting a cyber insurance product) and the models destined for insurers aid them with the design of cyber insurance products (including setting premiums) as well as with estimating risks (such as determining whether or not to issue a policy).

These models take a number of behavioural elements into account. They model the behavioural choices of various cyber threat actors in terms of their decisions as to whether or not to attack a company, thus progressing beyond the current CSRM frameworks that do not properly account for adversarial threats. They also consider the behavioural choices of companies and insurers, looking at companies' cybersecurity resource allocation and risk transfer decisions and insurers' risk assessment decisions. These models are presented in detail in Chapter 4.

In the next two sections, we provide background information on the key components of the cybersecurity risk management problem and on the current state of the cyber insurance market. This provides important context for the concepts presented in the rest of this book.

1.2   A schematic view of cybersecurity risk management

To provide the necessary foundation for many of the topics discussed in the rest of this book, this section gives an overview of the factors involved in CSRM. We present a basic schematic view in Figure 1.1. An *organisation* faces potential cyber *threats* that can have significant *impacts* upon it. It chooses what we call a *"cybersecurity portfolio,"* which consists of a combination of security products (security controls and recovery controls) and insurance products, that enables it to manage its cybersecurity risk as best as possible.

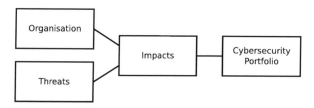

Figure 1.1: Basic schematic view of the factors involved in CSRM

Figure 1.2 elaborates upon the four elements in the basic schematic view. The *organisation* is described in terms of its *profile* and *assets*, together with what we shall call *other organisational features*.

With respect to *threats*, we use the Information Security Forum (ISF) classification of *targeted* cyber threats (in which the threat actors, or *attackers*, specifically aim their attacks at the organisation) and unintentional threats. The latter include *non-targeted* cyber threats (cyber attacks that are not directly targeted at an organisation, e.g. that are random or opportunistic), *accidental threats* (cyber incidents caused without malicious intent, e.g. due to human error or system failure), and *environmental threats* (natural disasters that are outside the control of the organisation, e.g. floods or earthquakes).

The impacts of an attack include *insurable* impacts, which can be partly covered by insurance products, and *non-insurable* ones, which are not covered by insurance.

The *cybersecurity portfolio* includes *security controls*, which are put in place by the organisation and consist of measures to prevent, protect against, and counter cyber attacks, including threat detection and response. This helps reduce the likelihood of threats occurring and mitigates their impact if they do. As in the ISF classification, we can further divide the security controls according to *procedural controls* (practices and procedures that enhance security), *technical controls* (technology solutions, e.g. software or hardware), and *physical controls* (physical measures, e.g. guards, gates, and security cameras).

In addition to security controls, the cybersecurity portfolio also includes *recovery controls*, typically implemented to respond to and recover from cyber attacks, reducing their impact. It also encompasses *insurance contracts* serving to transfer the risk, helping reduce the financial consequences of an attack. The above instruments will typically have to satisfy certain constraints when it comes to available cybersecurity budgets, compliance requirements, and so on.[1]

Each element in these boxes may involve a large number of components, which are detailed in cyber risk management catalogues. For example, the ISF's catalogue lists the the following tools to implement technical security controls:

- firewalls and internet gateways,

- secure configurations,

- access control systems,

- malware protection systems,

- backup systems,

---

[1]The business and investment concept of accepting risk, which occurs when a company believes that the potential loss from a risk is not significant enough to justify spending money to prevent it, is outside the scope of the schematic views.

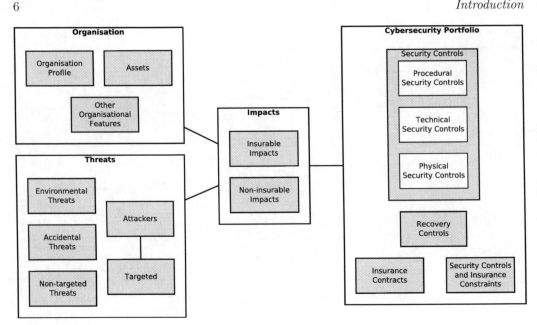

Figure 1.2: Detailed schematic view of the factors involved in CSRM

- intrusion detection systems,

- wireless access control systems,

- mobile device control systems,

- cryptographic solutions,

- DDoS protection systems, and

- other technical controls, which includes new controls that might be developed.

Note that the first four security controls are also part of the UK's Cyber Essentials. Moreover, for each of these security controls there could be a large number of potential providers, each with their own specifications.

Clearly, the many CSRM factors discussed above—the diversity of threats, the range of potential impacts, and the large number of products to choose from for the protection of cyber assets, from a plethora of security controls to relatively new products such as cyber insurance—illustrate the complexity of the CSRM challenges that must be overcome.

1.3 The current state of the cyber insurance market

This section provides an overview of the various facets of the cyber insurance market, setting forth basic concepts that later chapters build upon.

1.3.1 Decision-making within an organisation regarding the purchase of cyber insurance

When deciding upon its cybersecurity resource allocation and selecting a cyber insurance product, an organisation typically performs the following steps: It begins by considering its compliance requirements. Then it analyses its value-chain and classifies its assets. Next, the organisation assesses the existing security controls it has in place to protect its assets This is followed by an impact analysis of the effect of a cyber attack on its assets. If the organisation determines that a cyber attack could have a critical impact on operations, it will likely choose security controls and a cyber insurance product offering high levels of cybersecurity protection and financial coverage.

If it does not deem that the impact of an attack would be significant, the organisation will carry out a business continuity resilience evaluation to ensure that it can minimise the damage caused in the event of a cyber attack. The organisation will also carry out a financial impact quantification and cyber insurance price comparison, which involves simulating the financial cost of a cyber attack in certain high-risk scenarios and then for each cyber insurance product calculating a ratio of the cost of the premium to the amount of financial coverage provided by the cyber insurance product. If the ratio is greater than 1 (i.e. if the benefit of the cyber insurance product's financial coverage exceeds the cost of the premium), then the organisation will likely choose that cyber insurance product.

If not, given that:

1. the cost of the cyber insurance product would exceed the benefit of the cyber insurance product's financial coverage,

2. the organisation would not suffer a disastrous interruption from such an attack, and

3. the organisation is not subject to relevant regulatory or legal obligations,

then the organisation is unlikely to select the proposed cyber insurance product.

This cyber insurance decision-making process may be particularly difficult for SMEs, as they tend to have less cybersecurity knowledge and fewer resources to invest compared to larger companies.

1.3.2 Growth of the market

Although the cyber insurance market is still underdeveloped, it is growing steadily. The global size of the cyber insurance market was $3.89 billion in 2017 (Adroit Market Research, 2019) and is expected to reach $14 billion by 2022 (Sharma, 2018). The US is by far the largest market, accounting for some 90% of the total (EIOPA, 2018). Europe comes in second with a market size of €295 million and is expanding rapidly, having grown 72% between 2017 and 2018, the last year for which data was available (EIOPA, 2019).

Factors driving the growth in demand include greater awareness of cyber risks due to the rising number of attacks, highly publicised data breach incidents and increased losses associated with them, as well as fears of reputational damage caused by an attack (Inpoint, 2017; P&S Intelligence, 2017). The implementation and enforcement of cyber and information security-related legislation is also a factor. In the United States, the introduction of a number of data protection laws, including the 1996 Health Insurance Portability and Accountability Act (HIPAA), played a major role in increasing the demand for cyber insurance. More recently, the 2018 General Data Protection Regulation (GDPR) appears to be having a similar effect in the European Union. The £183 million fine levied upon British Airways under GDPR for its 2019 data breach was an unprecedentedly large sum and made companies operating in Europe take notice. While demand for cyber insurance in recent years has been led by large corporations from a wide variety of sectors that store PII, this seems to be changing as smaller companies become more aware of cyber threats and the costs associated with them (Inpoint, 2017; Aon Inpoint, 2018).

The supply side of the market is dominated by a few big players, although here too smaller insurers are increasingly offering cyber insurance products. In the US more than 60% of standalone premiums were written by the five largest insurance companies and in Europe the top three companies wrote over 70% of premiums (Inpoint, 2017). However, the number of insurers underwriting cyber insurance products is steadily rising. There has been a significant increase in insurers' appetite for cyber risk over the past decade, leading to an upturn in the number and variety of policies issued.

1.3.3   Types of insurance products

Cyber insurance products are focused on commercial business at present (EIOPA, 2018). They can be offered as a standalone product or added on to an existing policy as part of a package. The main types of coverage offered include business interruption as well as data loss and recovery, in addition to extortion, third party data breaches, and reputational loss. Policies generally cover the costs of repairing IT systems, carrying out forensic invest-igations, as well as notifying customers of data breaches and providing them with credit monitoring. In some cases, they may cover the cost of the fine for a data breach, but regu-lators do not always allow this to be covered by insurance. Insurance cover may also include services such as legal support, public relations, and crisis management.

Additional descriptions of the potentially insurable impacts, including the distinction between first party and third party impacts as with traditional insurance products, are available in Couce-Vieira et al. (2020a). Since the Target attack, a number of companies now require their third party suppliers to obtain separate liability coverage, as a condition of doing business with them.

Cyber insurance cover for individuals is also starting to emerge, given the potential for credit card or identity theft when using the internet and the increasing uptake of IoT devices in the home.

1.3.4   Challenges for insurers

Insurers face a multitude of challenges—many stemming from the difficulty of accurately assessing cybersecurity risk—that inhibit the development and accurate pricing of cyber insurance products. One of the biggest challenges that insurers face is the constantly evolving nature of cyber risk. As discussed earlier, the threats impacting organisations are growing in size and sophistication, with new threat actors and types of attacks regularly emerging on the scene. In tandem, the security posture of organisations can evolve rapidly as well. They may regularly take on additional third party suppliers, thereby increasing their cyber exposure, or implement new security products, decreasing it.

The challenge is compounded by correlations between various risks. Accumulation risk, or the risk of a claim from a single incident spreading to multiple lines of business, is a major concern for insurers. For example, a cyber attack that takes down the power grid will likely impact a wide range of sectors that depend upon it from transport to communications, affecting not just cyber insurance but also a host of other business lines including property, health, and even life insurance (Lloyd's and University of Cambridge Centre for Risk Studies, 2015). An even greater preoccupation for insurers is systemic risk, or the possibility that one incident could cause a cascading failure that triggers a collapse of the entire system[2] (Ducos and de Ligniéres, 2019). For example, a cyber attack that takes down the power grid could seriously threaten the viability of an insurance company or even of the insurance industry as a whole if its impacts are consequential enough.

Another challenge is information asymmetry, given that organisations seeking to purchase cyber insurance typically have more information about their risk posture than insurers. This relates to moral hazard, or the risk that an organisation engages in riskier behaviour because it has been insured. It can also lead to adverse selection, where an insurer provides insurance coverage to an organisation whose risk is much higher than the insurer is aware of. This results in an adverse effect for the insurer because it has issued an insurance policy at a cost lower than it would charge if it were aware of the actual risk, exposing the insurer to potential loses. Adverse selection is often due to an organisation seeking insurance coverage providing false information or withholding pertinent information from the insurer.

Insurers are also confronted with the lack of data (Anderson and Fuloria, 2010) when it comes to cyber incidents. Unlike in other domains that have similarly elevated levels of risk—such as pandemics—there is an absence of historical data to draw on when setting premiums. Organisations are also reluctant to disclose intrusion attempts or consequences of attacks due to reputational concerns, as this could negatively affect their relations with stakeholders or cause them to lose customers (Couce-Vieira et al., 2020a; Balchanos, 2012). This challenge is exacerbated by an acute shortage of experienced cybersecurity underwriters. Many underwriters have little knowledge of or experience with cybersecurity.

It can also be difficult to assess the financial impact of an attack that has occurred. For example, for insurance products covering reputational loss from data breaches, it can be particularly difficult to quantify the financial losses and implications for future revenues, i.e. whether the loss is permanent or temporary.

If insurers can overcome these challenges, however, this represents an opportunity for them to innovate in their offerings to help companies manage various degrees of risk.

1.4 The way forward

Given the many challenges that both companies and insurers face when it comes to CSRM, there is a vital need for innovative new approaches to risk management in cybersecurity for both customers and insurers. This book contributes to more rigorous methods of risk management in cybersecurity by:

- promoting cyber insurance as a key part of CSRM;

- developing new models that consider the behavioural aspects of attackers, overcoming a key issue with current CSRM frameworks that do not take the intentionality of threats into account;

- examining other behavioural aspects by factoring in the decisions of companies and insurers;

- developing a method of dynamically pricing cyber insurance products in response to changes in a company's cybersecurity risk profile, helping insurers deal with the rapidly evolving nature of cybersecurity risk;

---

[2]as opposed to the damage being contained to harming just one component of that system

- proposing a method to better quantify accumulation risk, enabling insurers to better manage correlated risks;

- proposing policy solutions to overcome moral hazard issues, assisting insurers with challenges linked to information asymmetry;

- using structured expert judgement elicitation techniques in instances where there is limited data, as a way to partially overcome the lack of data in the cybersecurity domain;

- facilitating the adoption of the models we have developed by designing and making publicly available a toolbox that provides an online interface for the model aimed at companies, as well as by providing a full case study illustrating how to implement the model; and

- giving particular consideration to SMEs, given that they may be especially vulnerable to cyber attacks.

These topics are further developed and expanded upon in subsequent chapters. Chapter 2 provides an overview of the cyber insurance ecosystem and examines the decision-making problems that organisations and insurers must contend with regarding risk management for cybersecurity and cyber insurance, drawing on psychological perspectives. It also makes use of an Agent-Based Model to assess the effects of various policy interventions on the ecosystem. Chapter 3 discusses the cybersecurity challenges that organisations face, and applies psychological and behavioural economics insights involving human behaviour and decision-making to cybersecurity and cyber insurance. It also uses Behavioural Economics Experiments to investigate the effects of behavioural interventions on cyber insurance uptake. Building on this, Chapter 4 presents a series of models to assist organisations and insurers with their decisions involving risk management in cybersecurity, including a key model to help organisations determine their optimal allocation of cybersecurity resources and select a cyber insurance product. It also provides auxiliary models to aid insurance companies with their risk management issues, enabling better quantification of accumulation risk and improved methods of designing and issuing cyber insurance products, including ways to dynamically price these products. Chapter 5 presents a case study to illustrate how to implement the key model developed in Chapter 4, providing detailed numerical examples. We conclude with a final discussion regarding the main points of this book and their implications for policy.

2

The Cyber Insurance Landscape

Katsiaryna Labunets, Wolter Pieters, Michel van Eeten
Delft University of Technology

Dawn Branley-Bell, Lynne Coventry, Pam Briggs
Northumbria University

Inés Martínez, Jhoties Sewnandan
Delft University of Technology

CONTENTS

This chapter provides an analysis of the cyber insurance landscape. We start with an overview of the cyber insurance ecosystem, presenting the cyber insurance life cycle and the main and secondary actors. We then examine organisational decision-making involving cybersecurity and cyber insurance, both for companies as a whole and for Small and Medium Enterprises in particular. We present the findings of studies we have conducted on these topics, which draw on the Burke and Litwin Performance and Change Model and on Protection Motivation Theory, and consider the policy implications. Finally, we simulate the effects of different policy interventions on the ecosystem, making use of an Agent-Based Model to do so.

2.1 The cyber insurance ecosystem

Cyber insurance is a risk transfer option that companies can use to mitigate IT-related losses in exchange for a premium (Fauntleroy et al., 2015). Product offerings mainly cover losses due to a virus or hacking or data breaches, but not IT incidents in general. In this section, we provide an overview of the cyber insurance ecosystem, including the cyber insurance life cycle and the main and secondary actors involved in cyber insurance processes. It is based on the cyber insurance and security risk management literature, an analysis of major cyber insurance products available on the market, and interviews with insurers.

2.1.1   The cyber insurance life cycle

The cyber insurance life cycle includes several processes. It begins with a request for an insurance offer. A company decides to buy cyber insurance to cover certain residual cyber risks and requests a price quote from an insurance provider (insurer). Next, the insurer assesses the company's risks. The risk assessment approaches used depend on the type of company (financial company, large corporation, etc.) and the internal workflow of the insurer. After completing the risk assessment, the insurer prepares a proposed contract, negotiations ensue between the insurer and the company, and once both parties agree the contract is signed.

The claims handling process is triggered when the insured company experiences an actual or a suspected data breach or cyber attack. The insured company must notify an incident manager and activate the company's insurance policy. Several parties may be involved in this process, e.g. a breach counsel for problem handling, a forensic investigator, and in the case of some incidents the regulator must be notified as well. Based on the information collected and the insurance terms, the insurer determines the payout.

2.1.2   Actors and relationships

Analysing this cyber insurance life cycle, we have identified the relevant parties and existing relationships between them, which we have modelled in Figure 2.1. Below we briefly describe the main and secondary actors in the model and their relationships.

The main actors are:

- An *insurance provider (insurer)* takes on companies' risk, providing them with appropriate coverage in exchange for payment (Marotta et al., 2017). Insurers aim to build a profitable portfolio of clients and increase their market share. They also look to build cybersecurity awareness among the companies they have insured.

- A *company* is a business that seeks to purchase cyber insurance. For the purposes of this book, we classify companies according to the following categories[1]:

    1. *Small and Medium Enterprises (SMEs)* have a staff headcount below 250 persons and a turnover below €50 million (European Commission, 2020).
    2. *Medium-sized companies* have a staff headcount below 2,000 persons and a turnover ≤ €500 million (Hargrave, 2019).
    3. *Large companies* have a staff headcount of more than 2,000 persons or a turnover above €500 million.

A company that purchases cyber insurance becomes an *insured company*. Insured companies' aims are to acquire a better picture of the cyber threats impacting them, transfer cyber risk-related losses in exchange for reasonable premiums, get advice on cybersecurity protective measures,[2] and receive assistance in the cyber incident response process.

- An *insurance broker or other intermediary* aims to provide advice to assist the company in the selection of an appropriate cyber insurance product. Brokers aim to run a profitable business and improve their market penetration. There are two types of brokers: retail and wholesale. Since for the purposes of this model we are primarily interested in brokers that sell cyber insurance products to companies, we only consider retail ones, who buy insurance products from either insurers or wholesale brokers and sell them to businesses

---

[1]Note that these categories will vary according to the insurer. Some insurers also have a *Jumbo* category.
[2]Not every insurer offers these services.

or individuals.[3] Other intermediaries that sell insurance products to companies include agents, who are employed by insurers to sell insurance products on their behalf, and insurance consultants.

The secondary actors, presented in alphabetical order, are:

- A *consumer* uses products and services provided by the company (Mentzer et al., 2001). Consumers aim to obtain a product or service that meets their needs in exchange for a reasonable cost. The interests of consumers are usually communicated to policymakers through consumer protection authorities, like the European Insurance and Occupational Pensions Authority (EIOPA), and by sharing the findings of research that has been undertaken.

- An *expert* provides assistance to the insurer in such areas as risk assessment, forensics, legal issues, public relations, and other services that the insurer may need. Experts can be either in-house or external.

- A *policymaker* takes part in the policymaking process. Policymakers might be national governments, members of parliament, and government agencies/public bodies, among others. They rely on information from varied sources such as research, representatives of special interest groups, associations of insurance providers or SMEs, and others. Policymakers' work results in the development of new policies, which a regulator is then responsible for implementing. In our context, the policymakers' goals involve raising the overall level of cybersecurity in the ecosystem. They also consist of managing the cyber risk to systems within their responsibility. To do so, policymakers might like to, for example, encourage companies to adopt essential security controls via cyber insurance policy requirements.

- A *reinsurer (reinsurance provider)* accepts a portion of the risk of the insurers' portfolio in exchange for payment. Reinsurance is essentially insurance for insurers (Kesan and Hayes, 2017). Reinsurers can be either private companies (e.g. Munich Re[4]) or governmental organisations (Robinson, 2012).

- A *regulator* is a public body responsible for the supervision and oversight of a particular industry or business activity, ensuring that organisations within that sector comply with all applicable laws, rules, and regulations (Levi-Faur, 2011). Some organisations are regulated by multiple regulators, depending on their business domain(s). Cyber insurance-related activities fall under the authority of a number of different regulators. This includes *insurance regulators* as well as *cybersecurity sector regulators*, such as national Data Protection Authorities (e.g. the "Commission nationale de l'informatique et des libertés" is the French Data Protection Authority). EU National Data Protection Authorities are also responsible for implementing EU regulations involving information privacy, such as the General Data Protection Regulation (GDPR), within their countries.

- A *researcher (research)* investigates cyber insurance-related topics. This might involve research conducted at universities or think tanks, in private companies, consultancies, or elsewhere. Research findings can inform the work of policymakers, helping evaluate the effect of policy interventions on the ecosystem. They can also assist insurers in developing cyber insurance policies.

- A *security provider* provides security products and services to other parties to safeguard their assets. Security providers can work directly with companies or by cooperating with insurers.

---

[3]By contrast, wholesale brokers act as an intermediary between retail brokers and insurers.
[4]https://www.munichre.com

- A *threat actor (threat)* is often a malicious actor that aims to launch an attack against the company (Kissel, 2013). There are also non-malicious threat actors that unintentionally cause harm, e.g. due to committing an error. In the cyber insurance context, the motivations and aims of threat actors could be essential factors for understanding the behavioural elements of an attack. It is also important to distinguish between insider threats and external ones. Insiders already have access to the company's information systems, while external threat actors need to obtain access in order to perpetrate an attack.

- A *vendor* provides companies with a product or service. This would typically be an equipment vendor supporting the company's businesses processes, such as a software provider or network provider (Mentzer et al., 2001). Companies may require their vendors to have cyber insurance and/or prove that they are compliant with cybersecurity regulations. [NB: Although technically speaking a security provider (listed above) is a type of vendor, we have separated these into two categories to emphasise an important distinction: a vendor covers all vendors providing the company with products or services, while a security company only provides the company with security products and services.]

2.2   The cyber insurance adoption process and its challenges

Now that we have established a clear picture of the cyber insurance ecosystem, we seek to better understand how cybersecurity and cyber insurance decisions are taken within this ecosystem. This section focuses on what happens at the company level. Section 2.2.1 examines organisational decision-making involving cybersecurity and cyber insurance within all types of companies. Section 2.2.2 then does a deep dive on the decision-making process within SMEs specifically. In each instance, we consider the policy implications.

2.2.1   General company-level decision-making process for cybersecurity and cyber insurance

Companies must make key investment decisions concerning cybersecurity measures (including cyber insurance) on a regular basis, but there is a lack of research directly investigating how companies make these decisions, as identified by Weishäupl et al. (2018). In particular, a recent literature review by Heidt et al. (2019) highlighted the scarcity of studies analysing IT-related security decision-making that take contextual factors into account, notably behavioural, environmental, and organisational ones. They found that these contextual factors are often overlooked because the majority of research in this area is quantitative in nature. They thus argue that it is important for research to consider such contextual factors (Heidt et al., 2019, p. 6145).

Study design

We drew on the Burke and Litwin (1992) Performance and Change Model, shown in Figure 2.2, in order to examine the drivers of IT-related decision-making, including the role of the contextual factors mentioned above. The Burke and Litwin Model is a general model describing the many factors that drive change within an organisation and serves as a useful starting point. The model illustrates how behaviour within companies can be influenced by a complex system of twelve factors. All the pathways between the factors are bidirectional, and therefore all factors, from company *structure* to *motivation* in the workplace, can feed into

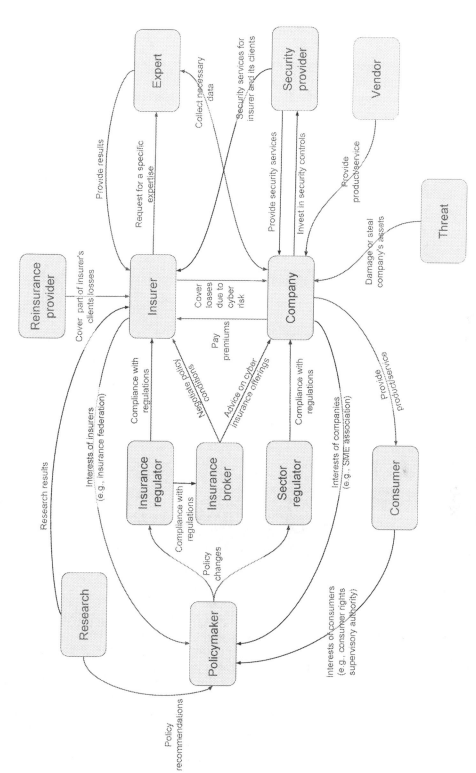

Figure 2.1: The cyber insurance ecosystem

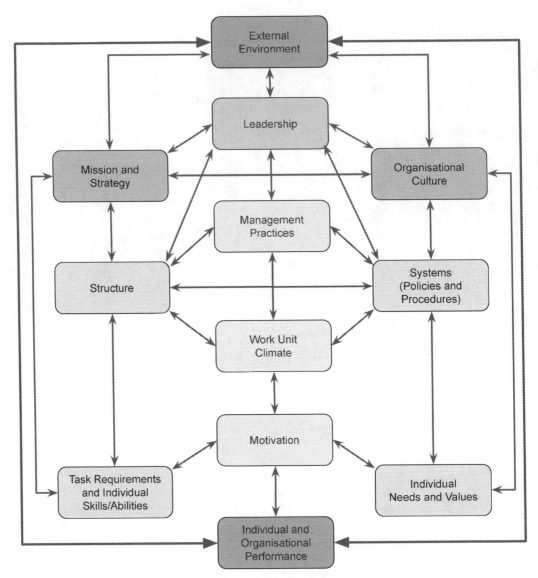

Figure 2.2: Burke and Litwin Performance and Change Model (adapted from Burke and Litwin, 1992)

organisational change in many different ways. The model ranks them in order of influence, with the most important factors at the top. The *external environment* is therefore the dominant factor in the model, having a significant impact on a company's *mission and strategy*, *organisational culture*, and *leadership*, and through them, on the other factors as well.

We can apply this model in a cybersecurity context. In order to identify the key roles and influential drivers of cybersecurity and cyber insurance specifically, we conducted 11 in-depth interviews with practitioners inside companies. This included individuals responsible for making cybersecurity decisions within a company as well as those involved in the sale/marketing of cybersecurity-related products and services (including cyber insurance). Those interviewed were from companies of different sizes, a mix of larger companies and

SMEs. We then carried out a qualitative analysis to identify and understand the influential drivers of cybersecurity-related decision-making within companies at board and senior management level.

Findings

We found that the decision-making process at company level involves a complex ecosystem in its own right. These systems can vary dramatically between companies, depending upon size, maturity, and sector. There is no universal 'one size fits all' structure for cybersecurity and cyber insurance decision-making within companies. There are also many different factors, both internal and external, that can influence companies' cybersecurity decision-making and cyber insurance adoption. Any cybersecurity services, products, and interventions need to account for this variation between companies in the decision-making process.

Internal drivers

There are many different processes influencing cybersecurity-related decisions inside a company. For example, cyber insurance adoption often seems to be driven outside of the technical teams (for example, from finance). Companies often have complex (and non-universal) structures involving numerous boards, committees, teams, and departments, each reflecting their own motivations, priorities, and ways of doing things.

In keeping with Weishäupl et al. (2018), we found evidence that companies can perceive cybersecurity-related decision-making (and related processes) to be time-consuming and tedious. For example, even the process of acquiring an insurance quote (and gathering the associated company information needed to obtain it) and the renewal process are seen as effortful. This can have a detrimental impact upon cyber insurance adoption, and is further compounded by a lack of awareness around cyber risk and cyber insurance coverage. Companies also expressed a mistrust of insurers, with concerns in regards to lack of transparency surrounding coverage. Resource and financial constraints also play a role.

External drivers

Cyber insurance adoption appears to be largely influenced by legislation and other policy aspects. In keeping with Weishäupl et al. (2018), our findings suggest that there may be a disconnect between the existing academic literature that sometimes regards cybersecurity decision-making as intrinsically motivated, and the emerging literature (such as this current study) that shows that companies may be more motivated to invest in cybersecurity because they need to do so to comply with legislation.

Legislation as a driver for cyber insurance also fits within the Burke and Litwin Model. As previously mentioned, this model suggests that the most dominant influence on organisational performance and change is the external environment. This could include factors such as legislation (e.g. the introduction of the GDPR) and media coverage of cyber risk—both of which were mentioned by those we interviewed as drivers of cyber insurance uptake. Therefore, in much the same way as Burke and Litwin, we found that external factors appear to have a strong influence on cybersecurity decision-making within companies.

Many approaches to cybersecurity assume a rational decision-making process. However, human decision-making and perception of risk does not always follow rational processes (Evans, 2003). This will be discussed further in Chapter 3. Many approaches also assume accurate calculations of benefit and risk—but this is unlikely at best, due to the current lack of data on cyber risk and how to measure it (Eling and Schnell, 2016).

Our findings suggest that companies may be responsive to more detailed cyber insurance policy wording regarding the specific terms and conditions of coverage (e.g. inclusions and exclusions). However, greater precision can make it difficult for policies to take into account the changing nature of the cybersecurity environment. Therefore a balance is needed between providing enough detail to reassure and/or guide companies, whilst maintaining enough room for policies to take into account new developments in cybersecurity risk and protection. Further research is required to investigate the most appropriate level of specificity. Legislation surrounding the standardisation of cyber insurance policy wording could help to reassure companies, and also address confusion over what policies cover (and clarify the perceived 'grey area' between traditional insurance policies and cyber policies).

Given companies' lack of confidence in insurers, policymakers should foster practices that could help build trust between insurers and insured companies. To achieve greater awareness around cyber risk and improve cybersecurity practices, policymakers can help partially overcome the issues involving the absence of good cyber incident data by promoting greater information sharing. There is a need for further investigation into the most appropriate ways to implement this.

2.2.2   Decision-making process in SMEs

Of the limited research that has been done on how organisations make decisions about cybersecurity and cyber insurance, it tends to focus on large companies. Yet, as the Burke and Litwin Model implies, the complexity of a company's ecosystem varies significantly according to a company's size. The cyber insurance adoption process for SMEs will therefore have a number of distinct characteristics. Moreover, SMEs face particular cybersecurity challenges. Relatively speaking they have fewer resources to invest in cybersecurity, making them more susceptible to cyber attacks. Research on this under-investigated sector is thus essential.

Study design

To investigate what mechanisms and factors influence how SMEs decide on cyber insurance adoption, we drew on the Protection Motivation Theory (PMT) Model. We use Rogers' 1983 revision of the model, which is the most commonly used version, shown in Figure 2.3 (Rogers, 1983; Floyd et al., 2000). PMT is a behavioural theory that identifies the elements that a decision-maker relies on to determine whether or not to protect against a threat. According to PMT, the *protection motivation* of an SME is based on three main components: *sources of information*, *threat appraisal*, and *coping appraisal*. There are two types of sources of information: *environmental* and *intrapersonal*. Environmental sources of information (*rewards*, *severity*, and *vulnerability*) are used to create the threat appraisal, and intrapersonal sources of information (*response efficacy*, *self-efficacy*, and *response costs*) are used to create the coping appraisal.

We conducted semi-structured interviews with representatives of ten SMEs. The interviews used semi-structured interview questions based on PMT. (The semi-structured interview guide is available in Labunets et al. (2019).)

Findings

We developed a Cyber Insurance Adoption Model for SMEs, shown in Figure 2.4, to illustrate an SME's decision-making process regarding cyber insurance (Martinez Bustamante, 2018). In the centre of the model is *cyber insurance adoption*. The other components show the cognitive process that a decision-maker uses to decide whether or not to purchase cyber insurance. It is based on the three central components of the PMT model—*sources of information*, *threat appraisal*, and *coping appraisal*—but extends it with two additional components: potential *impediments* and *drivers* of cyber insurance adoption. These two

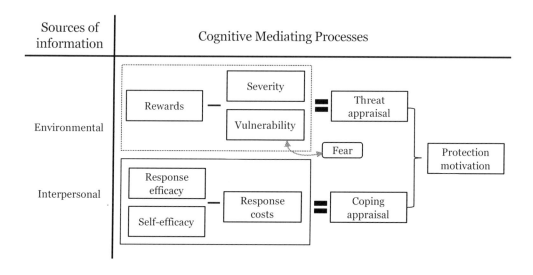

Figure 2.3: Protection Motivation Theory (PMT) Model (Rogers, 1983 revision)

additional components arose out of the interview process as important elements to consider as well.

Overall, the interview findings revealed that the cyber insurance decision-making process is problematic for SMEs due to poor understanding of cybersecurity risks and the dynamic nature of those risks. We provide more specifics about each model component.

Sources of information

We identified the following needs surrounding a company's *sources of information*: Cyber insurance is a relatively new concept for SMEs, so insurers and brokers have to be proactive in raising SMEs' awareness of cyber risks and available cyber insurance products. A policy measure regulating the role and liability of insurers and brokers in advising their clients on cybersecurity could benefit SMEs, as it is key for them to receive high-quality advice and trust their advisors. Insurers and brokers also need to be fully cognizant of the responsibility that they bear as advisors.

Another factor is that SMEs view the cyber threats they face and mitigation tactics they use as sensitive topics. This could explain why SMEs are often not willing to share their cybersecurity methods, which likely slows down the diffusion of cybersecurity best practices among SMEs. Further policy measures establishing and/or promoting cybersecurity certification schemes for companies and raising cybersecurity awareness could help SMEs obtain a clearer picture of their cybersecurity readiness. Cihon et al. (2018) agree: "Regulation should clearly signal to firms that certification helps meet their cybersecurity 'duty of care', which, if a breach were to occur, would see firms enjoy better defence against tort liability and fines." (NB: A cybersecurity certification framework is underway in the EU.)

Threat appraisal

Digitalisation is important for a company's growth, but it also creates new threat vectors through which a company can be attacked. While embracing digitalisation (e.g. cloud technology), SMEs are simultaneously concerned as part of their *threat appraisal* that this makes them more vulnerable to experiencing a data leak—which could affect the company's reputation and cause clients to lose trust in them. This drives them to increase their security measures in order to protect their data. It also pushes them to consider cyber insurance as a practical risk transfer strategy. As mentioned previously, a policy measure encouraging the development of standard language for cyber insurance policies could help SMEs better understand what residual risks they can transfer with cyber insurance.

Coping appraisal

When it comes to a company's *coping appraisal*, our findings indicate that cyber insurance is an attractive option for SMEs to transfer the risk of potential losses from a cyber attack if the insurance policy is clear and the premium price is fair. Since cyber insurance also provides policy holders with complementary information regarding cyber threats and help with cyber incident management, this creates added value for SMEs because they tend to lack knowledgeable personnel to deal with incidents. (However, if a company has invested sufficiently in cybersecurity protection measures and has skilled staff, then it has less motivation to buy cyber insurance.)

Further policy measures/regulations that impose financial costs on companies that experience cyber incidents (e.g. cost of notifying affected organisations or individuals following a breach, or fines in the event of a breach attributable to non-compliance with regulations) will likely motivate them to consider cyber insurance options. Various insurers already offer services helping companies ensure that they are compliant with existing regulations. Again, standardising cyber insurance policy language will help.

Impediments

Many of the *impediments* to cyber insurance adoption are problems that arise when companies try to buy cyber insurance. The process can be seen as complicated and there is often confusion about what a policy's coverage entails—as well as doubts about whether it will actually pay out in the event of an incident. High premiums also have a dissuasive effect.

Another reason for SMEs not purchasing cyber insurance is that a number of them believe they have a low probability of being attacked. Some companies, notably IT companies, think that they already have sufficient cybersecurity measures in place, so purchasing cyber insurance does not provide them with any added value. Finally, some erroneously believe that they have transferred the risk to the security provider; one of the interviewees commented that "it's not necessary for us to have insurance because [the security provider] has taken care of the [risk]".

Drivers

The main *drivers* of purchasing cyber insurance for companies include wanting to protect their reputation, which they do through various risk mitigation strategies, cyber insurance being one of them. Sectorial regulators make some recommendations in this respect.

Increasing awareness and experience of cybersecurity incidents also drives decision makers towards cyber insurance adoption. As mentioned previously, the additional services provided by cyber insurance (e.g. cyber threat information or incident assistance) motivates companies to purchase policies as well. Finally, small company status is a driver of cyber insurance adoption, with small companies increasingly realising that cyber insurance can help them.

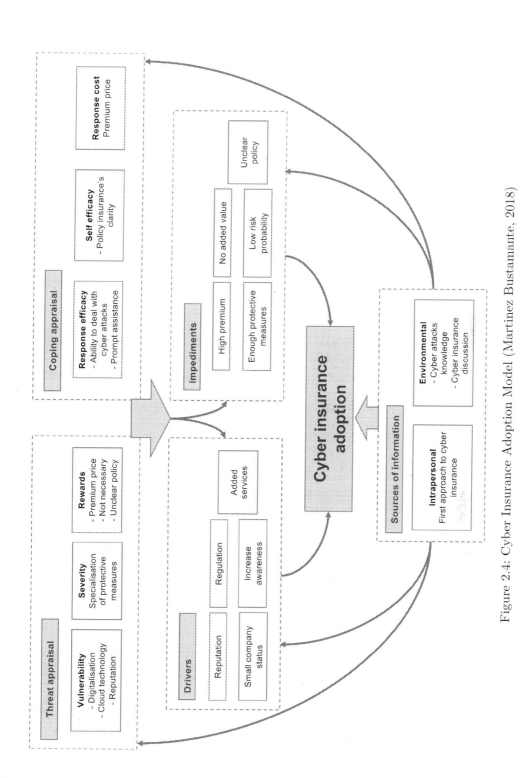

Figure 2.4: Cyber Insurance Adoption Model (Martinez Bustamante, 2018)

2.3   Effects of policy interventions

The previous section presented various factors impacting the cyber insurance adoption process. In this section, we discuss an Agent-Based Model (ABM) (van Dam et al., 2013) to simulate the effects of different types of policy interventions on overall risk. In agent-based modelling, system-level effects are studied based on simulating the behaviour of individual agents and their interactions.

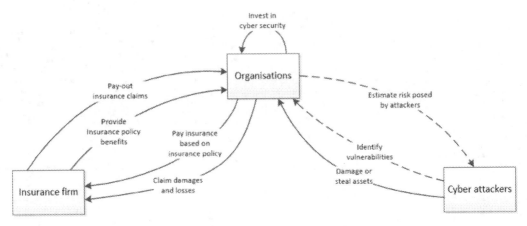

Figure 2.5: Simplified ecosystem for the Agent-Based Model (Sewnandan, 2018)

2.3.1   Study design

In order to keep the complexity of the model manageable, we used a simplified version of the cyber insurance ecosystem depicted in Figure 2.1 and its actors (i.e. the agents) as the basis for the model. This simplified ecosystem is shown in Figure 2.5 (Sewnandan, 2018).

For the agents in the simplified ecosystem, behavioural rules and parameters were defined as detailed in Sewnandan (2018) and the associated flow diagram is shown in Figure 2.6. At each "tick" of the model, representing a month in real time, agents observe their environment and execute actions. Security strength influences the risk of being attacked. Organisations conduct cyber risk management (right box), including making decisions about purchasing insurance and/or investing in improved security. Organisations also update their status (left box), including recovery from attacks and paying insurance premiums. Attackers can attack organisations (middle box), after which organisations can make insurance claims. Individual security strengths are updated through (a) effectiveness reduction over time, and (b) new security investments.

The ABM was implemented in NetLogo (Tisue and Wilensky, 2004), with an interface as shown in Figure 2.7. Sewnandan (2018) provides details on the parameters and the results of a sensitivity analysis, which examines how the uncertainty in the output of a model or system can be attributed to different sources of uncertainty in the inputs.

The system-level variables we use to study the effects of policy interventions are:

1. The *average security level* (on a scale of 0 from 1) in the ecosystem; and

2. The *global value loss* in the ecosystem (total asset value lost in euros, representing the inverse of *resilience*).

Using the model above, we investigated the effects of the following cyber insurance policy options on the ecosystem as a whole:

- *Package options*: the combination of the maximum amount in damages covered by the insurance and the insurance premium;

- *Contract length*: the duration of the insurance contract (6, 12, or 24 months);

- *Risk selection*: demanding improved cybersecurity levels,[5] or increasing the premium for clients when an insurer believes their cybersecurity levels need improvement;

- *Incentivisation*: lowering the premium for clients with high cybersecurity levels;

- *Upfront risk assessment*: requiring a potential client to perform a certain type of risk assessment first[6];

- *Sharing cybersecurity information*: providing clients with information on security controls, threats, etc. to help enhance their cybersecurity;

- *Requiring organisations to maintain their cybersecurity levels*: demanding that their initial cybersecurity levels are maintained to retain coverage.

We ran simulations for an ecosystem consisting of 125 organisations.

We also conducted a synergy experiment, which involves determining whether two or more discrete policy options can have a combined effect that is greater than the sum of the effects of each on their own. In essence, whether the whole is greater than the sum of its parts. In the experiment, we investigated the effects of combining the options *risk selection*, *incentivisation*, and *sharing cybersecurity control information*.

2.3.2 Findings

We measured the effect of the different policy options on (a) the average security level in the ecosystem, (b) the global value loss in the ecosystem (i.e. the total asset value lost, or the inverse of resilience), and (c) the percentage of insured organisations, under the model assumptions and parameter settings.

We observed that the effects of the different policy options on the average security level in the ecosystem are relatively small, with the synergy experiment providing the best results. For all policy options, the average security level was in the range of 0.54 to 0.58.

In terms of the impact on global value loss, the effects are small as well. In this case, the effect of the synergy experiment is somewhere in the middle compared to individual policy options. This suggests that although the combination of policy options improves overall security, it does not necessarily improve resilience, in the sense of reducing the global value loss. This could be because high-risk organisations might not purchase cyber insurance when the risk selection and incentivisation policy options are implemented, due to not being able to purchase it at an acceptable price.

Also, the synergy experiment results in a relatively low percentage of insured organisations (less than 40 out of 125 organisations, or 32%). This is because the combined policy options make cyber insurance less attractive for some (high risk) organisations, thereby reducing adoption but improving ecosystem-level security. The detailed overview of the results is available in Sewnandan (2018).

---

[5] In practice this would be among companies that already have reasonable cybersecurity levels, as insurers will not insure companies that have poor or low security levels. Insurers decline many risks based on a company having poor or low cybersecurity levels. The threshold will depend on each insurer's individual risk tolerance.

[6] At present, many insurers only assess a potential client's risk based on the client's application form.

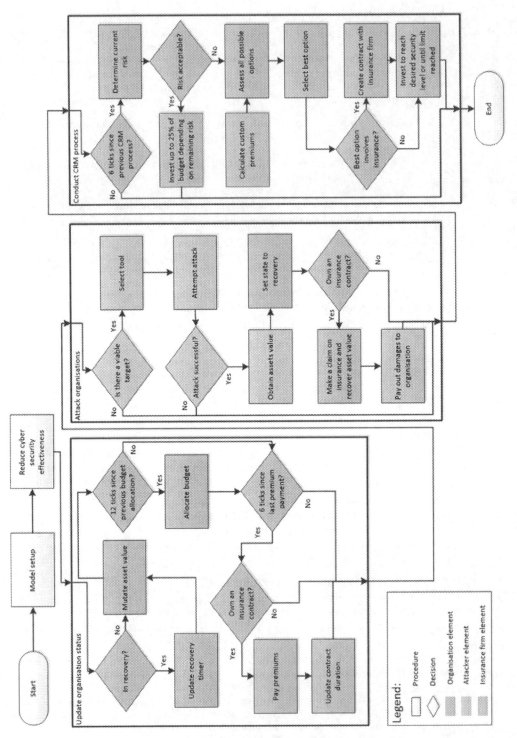

Figure 2.6: Flow diagram of the Agent-Based Model (Sewnandan, 2018)

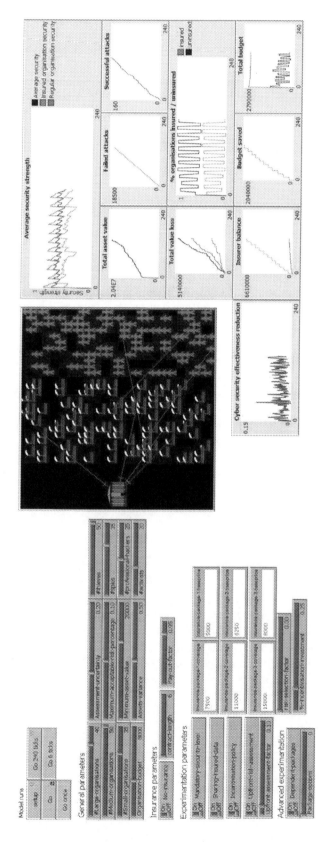

Figure 2.7: Interface of the Agent-Based Model in NetLogo (Sewnandan, 2018)

2.3.3   Discussion

Overall, we found that various individual insurance policy options had positive but rather small effects. The combination of several policy options into a synergetic design provided results with more observable effects at the ecosystem level.

More specifically, the following conclusions can be drawn from the Agent-Based Model and its results described above:

1. Under the assumptions in this experiment, the overall effect of individual policy options on the average security level and on resilience (as measured by global value loss) at the ecosystem level is small.

2. Combining different policy options results in a modest increase in the average security level but does not necessarily improve resilience, because high risk organisations may be discouraged from purchasing insurance.

3. Cyber insurance policy interventions can only have a large effect on the ecosystem in case of widespread adoption under the baseline condition. Some policy measures will actually be effective precisely because they reduce the number of insured organisations. For example, this might be by employing risk selection to avoid moral hazard (i.e. the risk that an insured company will engage in riskier behaviour because they have insurance).

As in any agent-based modelling exercise, assumptions had to be made regarding behavioural structures of the agents as well as model parameters. The model has been run with different variations of the parameters, and the insights above seem relatively robust. Nevertheless, further studies can investigate the effects of different assumptions on the ecosystem.

In terms of policy recommendations, we can derive that (a) policymakers should be aware that, depending on the circumstances, their key role may be in preventing negative impacts of cyber insurance rather than stimulating positive ones; and (b) policy measures that improve resilience may weaken overall security, because increasing the number of insured organisations may worsen the moral hazard problem. This trade-off is a key factor in decision-making.

2.4   Conclusions

In this chapter, we have presented the cyber insurance ecosystem. In Section 2.1, we discussed the different types of actors involved in the ecosystem and the relationships between them. In Section 2.2, we reviewed the cyber insurance adoption process in different types of companies. We also discussed the decision-making problems that companies, insurers, and insurance brokers may encounter in regard to cybersecurity risk management, as well as the factors influencing companies' decisions to purchase cyber insurance. Section 2.3 used agent-based modelling to simulate the effects of various policy intervention options on the overall risk in the ecosystem. These results shed light on the complexity of the system, the factors influencing the behaviour of some of the actors, and the effect of policy interventions on actor behaviour under certain assumptions. Further studies can enrich the findings by investigating the effects of different assumptions on the ecosystem and/or examining the behaviour of other actors to capture more of the complexity in simulations.

3

Behavioural Issues in Cybersecurity

Jose Vila

DevStat and University of Valencia

Pam Briggs, Dawn Branley-Bell

Northumbria University

Yolanda Gomez

DevStat

Lynne Coventry

Northumbria University

CONTENTS

This chapter opens with the challenges that organisations face in dealing with cybersecurity, then looks at the critical role of individuals' cybersecurity behaviour within an organisation in ensuring the cybersecurity of the organisation as a whole. We then turn to models of human behaviour and decision-making drawn from psychology and behavioural economics, examining their key insights for cybersecurity and cyber insurance. From psychology we consider the Theory of Planned Behaviour, Protection Motivation Theory, and others deriving from them, while from behavioural economics we consider Dual-Thinking Theory and Prospect Theory. We also look at Behavioural Economics Experiments, which help us investigate the effects of behavioural interventions, presenting an example involving cyber insurance. We conclude with a discussion of the benefits of combining psychological and behavioural economics approaches.

3.1   The cybersecurity challenge for organisations

Cybersecurity represents a large and growing business threat. The UK Government's *Cyber Security Breaches Survey 2019* found that 32% of businesses experienced at least one breach or attack in the past year (Finnerty et al., 2019). The estimated costs of such breaches vary widely, in part because while many breaches result in direct financial losses, it is difficult to account for secondary costs such as reputational damage. In addition, the majority of estimates are calculated by security and consultancy firms; therefore, biases may arise from their vested interests (Eling and Schnell, 2016).

The threats that organisations face are constantly evolving. For example, ransomware attacks have risen steadily in recent years, as noted in the European Union Agency for Cybersecurity's *Threat Landscape Report 2018* (ENISA, 2019). The past several years have also seen the growth of botnets that are more destructive than ever before, as they leverage the computing power of devices that are part of the burgeoning Internet of Things. Meanwhile, what we call "the human factor" remains unchanged: Regardless of the latest security products adopted by an organisation, employee error will always be a source of vulnerability, making it impossible to achieve 100% protection (Pal et al., 2017).

3.1.1   A role for cyber insurance in tackling the cybersecurity challenge?

This creates a market for cyber insurance. Cyber insurance policies are beginning to diversify but have tended to provide three basic types of coverage: (i) liability coverage in the event of a data breach, (ii) a means to remedy the breach, and (iii) support to repair reputational damage (Bandyopadhyay et al., 2009; Romanosky et al., 2019). An ideal scenario is that organisations would invest in both self-protection (e.g. firewalls and up-to-date antivirus software) and cyber insurance (Pal et al., 2017).

If widely adopted and well-functioning, cyber insurance has the potential to encourage market-based risk management for cybersecurity, with a mechanism for spreading risk among multiple stakeholders. It also has the ability to act as an incentive for organisations to invest in cybersecurity, which would reduce risk for the organisation investing and for their wider network. Uptake could also lead to data aggregation on best practices and better tools for assessing security, something that is currently lacking in relation to cyber insurance. We noted elsewhere that cyber insurance has the capacity to strengthen IT security for society as a whole (Baer and Parkinson, 2007; Kuru and Bayraktar, 2017).

However, despite the proposed benefits and the increasing risk of cyber attacks, uptake has not reached expectations. Low (2017) found that less than 10% of UK companies take out cyber insurance and two years later the *Cybersecurity Breaches Survey 2019* found that only 11% of businesses reported having cyber insurance (Finnerty et al., 2019). This number is considerably lower than would be expected, given the size of the cybersecurity threat.

3.1.2   Overview of key challenges in cybersecurity

It seems well established that many businesses fail to understand the threat from cybersecurity. Only 33% of businesses have cybersecurity policies in place and only 31% conducted a cyber risk assessment within the last 12 months (Finnerty et al., 2019). There is also a relatively low awareness of cybersecurity issues among employees, where inaccurate perceptions of risk can cause individuals and businesses to assume that cyberattacks will not happen to them (e.g. "my data is not interesting enough") (Eling and Schnell, 2016).

The problem may be particularly acute for SMEs. These may not possess the expertise or understanding to appreciate the risk to their business, such as the risk as a result of not having secured their data (Henson and Garfield, 2016). Advisen (2015) found that SMEs tend to consider that there is a low probability that they will experience a cyber attack and are thus less likely to engage with cyber insurance. These assumptions are dangerous as, contrary to popular belief, the majority of cyberattacks actually target SMEs (Meland et al., 2015).

Organisations are not investing sufficient time in understanding their vulnerabilities (Klahr et al., 2017) nor allocating adequate funding for cybersecurity (Fielder et al., 2016). There are some signs that this may be beginning to improve. Cybersecurity as an issue for the board is increasing: Nexus (2016) found that most organisations (82%) reported that their board of directors was "concerned" or "very concerned" about cybersecurity and information security. The *Cyber Security Breaches Survey 2019* also reported that 78% of businesses stated that cybersecurity was a "high" priority for their senior management, with 40% saying it was a "very high" priority (Finnerty et al., 2019), an increase compared to previous years. However, much still remains to be done.

Businesses, particularly SMEs, are often heavily restricted by the budget they have available for cybersecurity. Because of this, they are forced to make trade-offs regarding how they defend their systems (Fielder et al., 2016). The organisation also has to take into account both the direct costs of implementing a particular safeguard and the indirect costs that the safeguard may have on the business (e.g. a reduction in productivity, negative impact on morale, decline in system performance speed, or re-training costs) (Fielder et al., 2016).

A further, pressing issue is that most organisations still see cybersecurity as being the IT department's problem (Advisen, 2015; Eling and Schnell, 2016). This is unfortunate as it positions cybersecurity solely as a technical issue rather than as a business concern (Nexus, 2016), and it certainly overlooks the behavioural issues and biases that pervade cybersecurity decision-making.

3.2   Individual decision-making within an organisation

A significant factor behind poor cybersecurity within organisations is that individual employees behave in insecure ways, i.e. users' lack of "secure" behaviour may leave the company vulnerable to cyber attacks. In an analysis of security breaches reported across different sectors, 64% of incidents were judged to be "likely" due to human error (Evans et al., 2018) and humans are a major target for cyber attacks (Marinos and Lourenço, 2019).

Individuals are key stakeholders in the cybersecurity arena, not only as victims but also as defenders against attacks. However, not all individuals recognise or acknowledge their responsibility to protect the network. It thus becomes increasingly important to address the human component of cybersecurity within organisations and motivate the workforce to be better defenders. This is sometimes taken to mean that we should improve everyday computing practices, but the situation is more complex than this: We need to understand more about the way that managers and employees behave within an organisation when it comes to cybersecurity and the ways that we might encourage, or *nudge*, better behaviour.

Unfortunately, there is a lack of reliable behavioural data on users' cybersecurity decisions and actions (van Bavel et al., 2019). What users say they understand and do is not necessarily the same as what they actually understand and do. This is a problem for a literature that is heavily reliant upon survey methodology. Users may report awareness in surveys but might not have the skills or inclination to carry out associated actions. This poses a real challenge to researchers addressing the human component of cybersecurity decision-making in general and decision-making around cyber insurance in particular.

There is a stronger literature around *attitudes* towards cybersecurity (Bulgurcu et al., 2010). When questioned, individual employees often show a resistance towards complying with their organisations' information security policies and a negative attitude towards cyber insurance uptake. Factors that underpin such attitudes can be grouped into three categories: (i) *failure to admit the risk* or take responsibility for it, due to misplaced confidence; (ii) *overstretched staff resources*, leading employees to feel that cybersecurity procedures are overly burdensome; and (iii) the influence of the immediate *social environment* in cybersecurity decision-making. Each of these factors is elaborated upon below.

3.2.1   Failure to admit the risk

Employees can show a range of inaccurate perceptions of risk. This includes cognitive biases known to affect decision-making, including optimism bias (e.g. "It won't happen to me") and confirmation bias (e.g. "We haven't been attacked yet, so we are well protected"). This sense of not feeling personally at risk is sometimes accompanied by the over-optimistic belief that "only amateurs fall victims to attacks" (Sasse and Flechais, 2005; Pfleeger and Caputo, 2012). People can easily become complacent when it comes to risk. For example, Miyazaki and Fernandez (2001) found that those who spent more time on the internet tend to have lower levels of concern over privacy and fewer worries about online purchases. In the absence of adverse events, and therefore a lack of learned experience, consumers perceive online risks to be overstated and/or sensationalised by the media.

Users' beliefs about their susceptibility to an attack directly impact their motivation to behave securely. To illustrate this, Davinson and Sillence (2010) found that training interventions around phishing failed to improve secure behaviour unless people changed their views about their own vulnerability.

We find evidence that employees can be both overconfident and underconfident in their own ability to protect against cyberattacks. On the one hand, Furnell (2007) found that people have a tendency to overestimate their ability to behave securely online. In their study, respondents claimed to have a good awareness of cybersecurity threats and safeguards, yet there were a number of areas in which they left themselves vulnerable. The overconfidence reflects a control bias in the sense that people feel they can exercise more personal control over their environment than is justified by the data (e.g. "I could easily detect a threat" or "I wouldn't fall for a scam") (Pfleeger and Caputo, 2012). This is a concern, as the more confident people are, the more risky their cyber behaviour becomes (Campbell et al., 2011; Weinstein, 1980). This notion is reinforced by Aytes and Connolly (2004), who found that knowledge of security threats was a poor predictor of students' attempts to mitigate threats, as even those who were aware of the dangers still engaged in unsafe computing practices. They concluded that providing information through awareness training was therefore not sufficient in itself to change behaviour.

On the other hand, underconfidence links to a well-known literature around a lack of computer self-efficacy. In such instances, people express doubts in relation to their ability to comply with information security policies or worry about their lack of access (actual or perceived) to the necessary organisational resources, e.g. training, policies, etc. In other cases, people may simply lack the skills to protect themselves. Furman et al. (2011) found that users were aware of and concerned about online and computer security but lacked a complete skill set to protect their computer systems, identities, and information online, when measured through a comprehension test.

3.2.2 Overstretched staff resources

Employees often view cybersecurity as obstructive, burdensome, and interfering with their primary work (Beautement et al., 2016; Turland et al., 2015). In other words, they see cybersecurity as highly demanding of their already stretched resources. They simply do not have the time and energy to spend on cybersecurity measures and often feel that they are overly and sometimes unnecessarily burdensome. This can leave employees somehow feeling that cybersecurity measures are unfair on the workforce. Herath and Rao (2009) found that employees held less favourable views towards cybersecurity policies if they believed that compliance with those policies was costly; e.g. it hindered their day-to-day job activity. This can lead to a tendency to look for workarounds, especially if staff are unaware of how their behaviour compromises security (Sasse and Flechais, 2005).

Individuals weigh the costs and benefits of particular actions, and this holds true of security-related behaviours. We already know that a great deal of insecure online behaviour is simply driven by economic incentives. For example, users will happily ignore the security credentials of a website if it might be beneficial financially. Such motivators can lead to ill-advised downloads, use of insecure sites, and excessive disclosure of information. This can include being offered a favourable price (Kirlappos et al., 2012; White, 2004), a personalised service (Chellappa and Sin, 2005), prizes (Earp and Baumer, 2003), or a particularly desirable product (Kirlappos et al., 2012; White, 2004).

A deeper understanding of the ways in which behavioural incentives—both negative (*sanctions and punishment*) and positive (*rewards*)—can drive behaviour is useful here. Employees often believe that they will not be held accountable if they fail to follow the organisation's information security policy (Sasse and Flechais, 2005). However, employees will be more inclined to follow security policies if they feel that miscreants are likely to be caught. The important thing here is the likelihood of detection, as Herath and Rao (2009) found that people are motivated by the certainty of detection but not the severity of the penalty. Quite the opposite, rather surprisingly: Higher penalties are associated with poorer security behaviours, although the effect is modest. As the authors note, the security literature is inconsistent as regards the effects of sanctions (Kankanhalli et al., 2003; Pahnila et al., 2007) but seems to suggest that the existence and visibility of detection mechanisms is more important than the severity of the penalty.

Employees' perception of cybersecurity as burdensome and costly is exacerbated by most organisations' failure to reward good cybersecurity behaviour. Individuals are not usually recompensed for following best practice when it comes to cybersecurity (as opposed to meeting their main productivity goals).

3.2.3 Social environment

Other peoples' beliefs and behaviours help us to determine social norms, i.e. the unspoken 'rules' that indicate which behaviours are considered appropriate. These rules can vary between groups and reflect the local organisational culture. Leach (2003) believes that a strong security culture can be the best way to motivate staff to act in a secure manner, but this can backfire if others are not seen to be acting appropriately. In other words, social influence can have unexpected effects. Cialdini et al. (2006) warn of the dangers of telling staff that a particular (and inappropriate) behaviour is relatively widespread as this can act to reassure employees that they are behaving normatively.

Herath and Rao (2009) found that normative beliefs about the expectations and perceived behaviours of superiors, peers, and IT personnel had a strong impact on employees' security behaviour and their intentions to comply with security policies. In other words, we are strongly influenced by those around us. Indeed, organisational teams can come to develop their own local "shadow security" beliefs and expectations about what behaviours are required (Kirlappos et al., 2014). These are often different from the behaviour set forth within an information security policy, and are unlikely to be secure, having been modified by employees believing that they contain "unrealistic expectations" put on them by the security staff.

Another potentially damaging social effect is "social loafing" (George, 1992). This occurs in cases where individuals feel that others are carrying the load when it comes to cybersecurity protocols and so they need not bother. Stafford (2017) discusses "cyber loafing and cyber complacency" within organisational teams, noting that carelessness within those teams is often justified with reference to some kind of "higher power" that takes responsibility for cybersecurity, with examples including "No worries! I've got a Mac" or "The boss has my back" (pg. 9). This kind of thinking was also observed in Nicholson et al. (2018), who used their "cyber survival task" to identify mismatches in the thinking of IT staff and employees. They found that staff were relatively blasé about updates, believing (erroneously) that "if the update was important, it would get pushed through by the IT staff regardless" (pg. 434). The authors noted that employees often held the assumption that cybersecurity responsibilities were not their concern, but sat solely with members of the IT department.

3.3   Modelling pyschological and behavioural economics factors

The factors described in the previous section can be best understood with reference to a larger, long-standing literature that attempts to model human behaviour and decision-making in psychological or behavioural economics terms. These two disciplines have established approaches to conducting behavioural research in cybersecurity and cyber risk. Psychological models of behaviour typically recognise that behaviour is often planned and behavioural intention is typically influenced by attitudes towards that behaviour. They also recognise that these attitudes are influenced by our social environment (perceived "norms").

In contrast, behavioural economics models of behaviour typically acknowledge that decision-making is not entirely rational and people will show different biases in their decision-making in risky environments. They also discern that seemingly irrational behaviours can sometimes result when people are presented with overwhelming amounts of information or too many cognitive demands. In these circumstances, people might find shortcuts, or "heuristics", as a means to cope with this. We expand on the pyschological approaches in Section 3.4 and on the behavorial economics approaches in Section 3.5.

3.4   Psychological models

In this section we present several psychological models of human behaviour and decision-making and the key insights for cyber insurance and/or cybersecurity that can be obtained from them. We first introduce the Theory of Planned Behaviour (TPB), which is the foundation of Protection Motivation Theory (PMT) and a family of related models. We then briefly review PMT, which was already introduced in Chapter 2. It is the dominant model used for studying cybersecurity behaviour in the psychology literature. This is in part because the emphasis on threat appraisals and coping appraisals is particularly useful within

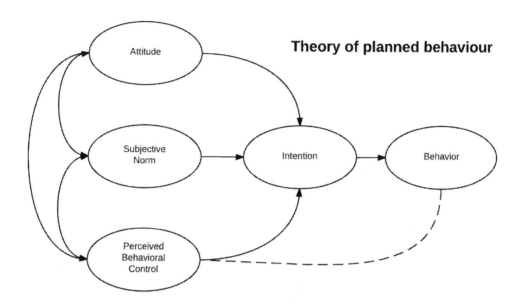

Figure 3.1: Theory of Planned Behaviour (Ajzen and Madden, 1986; Ajzen, 1991)

the cybersecurity context. We also present three additional models that build on Protection Motivation Theory: Herath and Rao (2009)'s combination of PMT and the TPB; the Extended Parallel Process Model (EPPM); and the Health Belief Model.

3.4.1 Theory of Planned Behaviour

As mentioned, many human behaviour and decision-making models derive from the Theory of Planned Behaviour (Ajzen and Madden, 1986; Ajzen, 1991), shown in Figure 3.1. TPB captures the ways that *attitudes* and beliefs drive behavioural *intentions* which, in turn, drive *behaviour*. Beliefs include both *subjective norms*, or an individual's belief that a key person or group of people will support their engaging in a particular behaviour, and *perceived behavioural control*, or an individual's belief in their own competence to engage in a particular behaviour.

In cyber insurance terms, this implies that encouraging people to adopt positive attitudes towards cyber insurance (e.g. increasing the perceived benefits compared to the perceived costs) could bolster their intentions to engage in better cybersecurity practices, potentially including cyber insurance uptake. In addition, raising normative beliefs (e.g. the belief that others think that cyber insurance is a good idea) is likely to improve cyber insurance adoption. To date, relatively little research on cyber insurance has used TPB as an explicit model, but it is commonly used as a model for predicting security compliance.

3.4.2 Protection Motivation Theory

Protection Motivation Theory is a model of risk assessment and behaviour change and, as noted, is derived from the Theory of Planned Behaviour and has been the primary psychological model used for studying cybersecurity behaviour. Since we already introduced PMT in Chapter 2, we will not reproduce it here but simply recapitulate its three central components: *sources of information*, *threat appraisal* (which is influenced by *severity* and *vulnerability*), and *coping appraisal* (which is influenced by *response efficacy* and *self-efficacy*).

According to PMT, when assessing a threat there are four factors that decision makers take into account and that drive their behaviour: (i) severity, or the perceived severity of the threat, (ii) vulnerability, or the perceived probability of its occurrence, (iii) response efficacy, or the perceived efficacy of the response to the threat, and (iv) self-efficacy, or decision makers' perception of their ability to respond effectively.

3.4.3   Protection Motivation Theory and the Theory of Planned Behaviour combined

Herath and Rao (2009) used a combination of Protection Motivation Theory and the Theory of Planned Behaviour to develop a predictive model of employee information security compliance intentions. More specifically, they used a set of constructs from PMT together with constructs from the TPB, and also incorporated factors known to directly affect organisational commitment. Based on this, they created a survey tool to examine the factors affecting employees' intentions to comply with their organisations' information security policies.

They found that, among PMT constructs, response efficacy and self-efficacy had a direct and significant impact on employee security compliance intentions and also found that social influence played an important role. In contrast, they found that response cost and security concern (i.e. concern about security threats) did not play a significant role in employee security compliance intentions. While this finding was surprising at the time, it now sits comfortably with a much larger set of studies that suggest that the coping appraisal component of PMT is the most important predictor of cybersecurity behaviour.

This suggests that behavioural interventions that use fear as a motivator are unlikely to be successful when used in isolation, which is interesting given the propensity of organisational and governmental cybersecurity campaigns to use threat imagery. The implications for cyber insurance are also important, as threat messages would seem less likely to drive cyber insurance uptake than messages about the importance of cyber insurance in driving response efficacy. Where threat messages are used, they should be accompanied by clear guidance on the relevance of available cyber insurance products.

3.4.4   The Extended Parallel Process Model

The Extended Parallel Process Model (EPPM) (Witte, 1996), illustrated in Figure 3.2, expands on PMT and explains how people appraise and respond to threats. EPPM proposes that perceived efficacy and perceived threat predict people's acceptance and willingness to act upon received threat messages. As in PMT, perceived efficacy has two dimensions: *response efficacy*, reflecting the belief that a particular response to a threat will be effective, and *self-efficacy*, reflecting the belief that the person can execute the necessary response effectively. Perceived threat also has two dimensions: *severity*, or perceived severity of the threat, and *susceptibility*, or perceived susceptibility to the threat. However, EPPM recognises that people typically exhibit two different types of response to a threat: *danger control response*, which involves controlling and responding to the threat, or *fear control response*, which involves controlling the fear they feel from the threat, often by engaging in denial processes, i.e. becoming resigned to the threat and/or downplaying it.

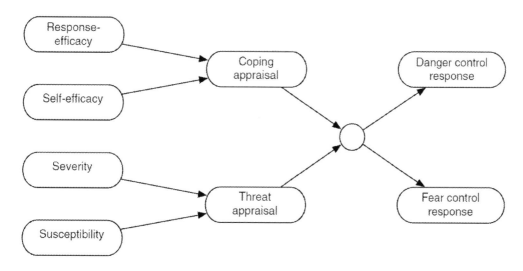

Figure 3.2: The Extended Parallel Process Model (Witte, 1996)

Again, we see implications here for the way that cyber risk is perceived. If an individual does not feel that they have the tools or the knowledge to cope with a cyber threat, then they may pretend it does not exist and give up on defending against it entirely.

The EPPM has been adopted to understand the way that emotional arousal associated with cybersecurity issues can act to influence an individual's assessment of a cybersecurity crisis (Zhang and Borden, 2019). In an experiment which examined how threat information (high vs. low) and behavioural recommendations (present vs. not present) impact individuals, the authors found that, when presented with "high" threat information (e.g. a severe data breach), participants who were given behavioural recommendations (i.e. specific steps to take to protect themselves) had higher anxiety than those who were not given behavioural recommendations, which at first seems surprising. The authors argued that anxiety in this case became an activating condition capable of driving behaviour change. In contrast, those without any behavioural recommendations in the face of a high threat simply felt sadness (a de-activating emotion) and resigned themselves to the cyber risk.

There is no doubt that threat messages are capable of inducing anxiety, but in EPPM terms, these threat messages do not always trigger behaviour that will reduce the threat. Instead, it appears that it is behavioural recommendations that can produce an impact. This is supported by Jansen and van Schaik (2019), who conducted a survey that found that strong fear appeals had marked effects on attitude, but no ultimate effects on self-reported behaviours. Similarly, and in perhaps the most unusual methodology employed to examine cybersecurity behaviour, Warkentin et al. (2016) used fMRI technology to understand the role that fear appeals have in determining behavioural intentions involving cybersecurity. They concluded that a focus on threats might be misplaced and that interventions that provide an actionable response to the threat (i.e. coping instructions) are more likely to prove successful as an intervention.

We should also note an extensive review of sixty years of fear appeal research that largely describes psychological models used in predicting health behaviour (Ruiter et al., 2014). The authors concluded that coping information aimed at increasing perceptions of response effectiveness and especially self-efficacy is more important in promoting protective action than presenting threatening health information intended to increase risk perceptions and generate fear arousal.

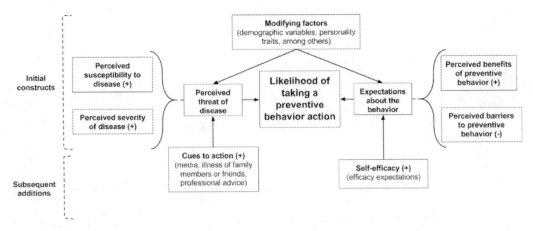

Figure 3.3: The Health Belief Model (as interpreted by Dodel and Mench, 2017)

3.4.5   The Health Belief Model

The Health Belief Model is also closely related to PMT and, while originally developed to model health behaviour, has been found to have important implications for the cybersecurity domain (Dodel and Mesch, 2017; Davinson and Sillence, 2010). Figure 3.3 illustrates the version adopted by Dodel and Mesch (2017). Beyond PMT elements, the Health Belief Model includes additional factors such as *perceived barriers to preventive behaviour*, *benefits of preventive behaviour* (e.g. through a cost/benefit analysis), and *cues to action* from the environment.

In the cybersecurity context, the Health Belief Model suggests users will take protective actions if: (i) they feel that a negative condition (e.g. computer virus) can be avoided and they also believe that (ii) the recommended action will be effective (e.g. doing a virus scan will prevent a viral infection) and that (iii) they can successfully perform the recommended action (e.g. the user is confident that he knows how to install virus protection files). The model also submits that users will only engage in cyber safety behaviours if the benefits (e.g. keeping the individual or the organisation safe) outweigh the costs (e.g. cost of antivirus or time required to download). The model also emphasises the importance of cues to action, suggesting that a key intervention would be to ensure that cyber risk messages are made personally relevant.

Dodel and Mesch (2017) employed the model to investigate users' likelihood of taking preventative action, based on data from a survey of Israeli internet users. They found that beliefs about digital threats and protective actions accounted for more variance in their choosing to adopt antivirus behaviours than socio-demographic characteristics or frequency of internet use. They also found that the coping elements of the Health Belief Model were key, concluding similarly to Davinson and Sillence (2010) that "Campaigns to reduce the perceived barriers to digital safety-related behaviours and strengthening the belief in the effectiveness of anti-malware seem to be clear and cost-efficient policy measures to increase the engagement in cyber safety" (pg. 366).

3.5   Behavioural economics models

In recent decades, new behavioural economics models have been proposed and are facilitating a deeper exploration and alternative interventions to improve cybersecurity behaviour. These formal models of human behaviour provide an important theoretical framework. Recognising that decision-making is not entirely rational, behavioural economics models go beyond the classical approach of perfect rationality and maximisation of expected utility. Section 3.5.1 presents key behavioural economics models that can be used to describe cybersecurity behaviour.

In addition, behavioural economics also provides a methodology based on the application of experimental methods to test these theories and generate empirical evidence as to how cybersecurity decisions are actually made. These include Behavioural Economics Experiments (BEEs), which are introduced in Section 3.5.2, alongside an explanation of how to conduct such experiments.

3.5.1   Models

In this section we present two key behavioural economics models that have been used to describe cybersecurity behaviour. First, we present Dual-Thinking Theory, a model for human decision-making set out by Nobel laureate Daniel Kahneman in his seminal book, *Thinking, Fast and Slow* (Kahneman, 2011). We then present Prospect Theory, which is a model to explain behavioural change that forms the cornerstone of behavioural economics. These models are both instrumental in order to understand BEEs.

Dual-Thinking Theory

Kahneman (2011) proposes a dual model for human decision-making that has important implications for decision-making involving cybersecurity. According to the model, all decisions (including of course those related to cybersecurity) are made employing two fundamentally different modes of thought, called System 1 and System 2. Roughly speaking, System 1 thinking is fast, intuitive, associative, metaphorical, automatic, and impressionistic, and cannot be switched off. Its operations involve no sense of intentional control. System 2 thinking is slow, conscious, deliberate, and effortful. System 2 thinking bears a close resemblance to the rational agent (termed *econ*) in standard economic theory, which considers decision makers as *econs* able to optimise their decision-making.

Prospect Theory

Prospect Theory (Kahneman and Tversky, 1979; Tversky and Kahneman, 1992) provides an economic model of behaviour under risk that proves especially useful for analysing cybersecurity behaviour (van Bavel et al., 2019). Prospect Theory departs from conventional economic models such as Expected Utility Theory in that: In Expected Utility Theory, a utility function u transforms objective outputs (for instance, monetary values) into their corresponding subjective values for the decision maker. Then, decision-making is determined by the optimisation of expected utility. The corresponding probabilities remain unchanged. As a simple example of how Expected Utility Theory works, let us consider that an agent is given the option to pay an amount I to participate in a game. In this game, she can obtain a net outcome x_i with a known probability p_i for $i = 1, ..., n$, where $x_1 > x_2 > \cdots > x_n$. The net outcome is equal to the total outcome of the game minus the participation cost I.

She will participate in the game if and only if $u(I)$ is lower than the expected utility of the outcomes given by $\sum_{i=1}^{n} p_i u(x_i)$.

By contrast, Prospect Theory considers that probabilities also need to be transformed before their consideration in the optimisation process. This transformation is done using the *weighting function*, denoted by w. The underlying idea is that, in the same way that an increase of €1,000 in the output does not increase the utility by the same amount if the initial output is €0 or €10 million, an increase of 0.10 in the probability has a different impact on the decision weight if it applies to a probability of 0.01 or 0.30. To capture this critical behavioural effect, w is defined in terms of probability *ranks*. A rank, or more intuitively a *good-news probability*, for any potential outcome x is defined as the probability of obtaining an outcome strictly larger than x. Formally, $rank(x) = \sum_{x_i > x} prob(x_i)$, and ranks are numbers between 0 and 1, where 0 is the rank associated with the best possible outcome and 1 is the rank associated with the worst.

Let us define $x_{n+1} = -\infty$. Then, the probability of outcome x_i can be written as $p_i = rank(x_{i+1}) - rank(x_i)$ for $i = 1, \ldots, n$. Given a weighting function w, the *decision weight* of outcome x_i is defined as $\pi_i = w(rank(x_{i+1})) - w(rank(x_i))$. Notice that if the weighting function is the identity function, i.e. $w(p) = p$, then the decision weights coincide with the probabilities of the outcomes, $\pi_i = p_i$. Decision weights are positive numbers lower than one, but they are not required to add up to one. Decision weights are related to the slope of the weighting function: the steeper the weighting function is, the larger the difference between $w(rank(x_{i+1}))$ and $w(rank(x_i))$ and then the larger the corresponding decision weight π_i. Under Prospect Theory, an agent with utility function $u(x)$ and weighting function $w(p)$ will participate in the game if and only if $u(I) < \sum_{i=1}^{n} \pi_i u(x_i)$.

Implications

Although a discussion comparing conventional and behavioural approaches to decision-making may seem too technical, it has relevant policy implications for cyber insurance. Indeed, Prospect Theory establishes that cyber insurance and cyber protection decisions are not made based on the actual risk of experiencing cyber attacks, as captured by the probabilities of suffering such attacks. Instead, these decisions are made in terms of decision weights.

Moreover, Prospect Theory provides the foundation for a methodology to estimate such decision weights in the form of BEEs. Note that this difference is not the result of agents having imperfect information regarding cyber risks, but of the psychological and cognitive mechanisms involving System 1 thinking. Decision weights do not coincide with risks, even if agents are informed and are able to accurately ascertain the value of those risks.

Two implications of the role of weighting functions in the field of cybersecurity are especially relevant. First, the decision as to whether or not to purchase a cyber insurance policy, and therefore the maximum premium that a potential client will pay for the policy, is conditioned by the shape of the weighting function. The estimation and calibration of this function, which can be performed using the Behavioural Economics Experiments presented in the next section, becomes a key tool to determine the optimal pricing of cyber insurance portfolios (Alventosa et al., 2016). Second, the critical role of the weighting function in driving cybersecurity behaviour provides an opportunity to design interventions aimed at enhancing cybersecurity by changing the shape of this function.

3.5.2 Behavioural Economics Experiments

Behavioural interventions, especially nudges addressing System 1, cannot be studied using conventional methods such as quantitative surveys or qualitative interviews (Hernández and Vila, 2014). Since System 1 thinking works in a subliminal manner, as humans we are not conscious of the role that an intervention may play and, for this reason, we cannot rationally elaborate on its potential impacts when answering a questionnaire or participating in a focus group. To know how System 1 thinking will react to an intervention, we need to apply a different methodology, specifically designed to assess System 1. This is what Behavioural Economics Experiments (BEEs) are designed for. In this section, we briefly present how BEEs work and how they can be used to study behavioural interventions for cyber insurance, illustrating this with specific examples.

Designing Behavioural Economics Experiments

Since humans cannot in general foresee the effect of a behavioural intervention on their cyber behaviour, and might even deny that such an influence takes place, the best way to establish the presence of and quantify such an effect is by direct observation of actual behaviour changes when the subject is exposed to an intervention. Moreover, we need to isolate the intervention from other stimuli to avoid confounding effects.

The scientific method, based on the application of controlled experiments to test research questions empirically, provides us with a useful tool to study behavioural interventions. The core of the method is the *random allocation* of each subject in the sample to either a treatment group (subjects affected by the intervention) or a control group (subjects not affected by the intervention). Then, assuming that the randomisation has created groups that can be compared, the differences in cybersecurity behaviour between subjects in the treatment and control groups could be attributed to the behavioural intervention applied to the treatment group. Of course, the ideal way to run such experiments would be to do so in a real-world situation, in a *field experiment*. However, the implementation of behavioural field experiments raises significant logistical and ethical concerns which, in general, limit their application in cybersecurity research.

BEEs provide a more feasible alternative to observe actual cybersecurity behaviour under controlled conditions, facilitated by the random allocation to treatment and control groups. As in any experiment, a BEE is an orderly procedure carried out with the goal of verifying or refuting the hypothesis that the intervention does affect subjects' behaviour. One of the main elements of a BEE is the *baseline experiment*, that is, a controlled, gamified environment where the subjects' behaviour will be observed. A key part of the baseline experiment is the instructions informing the subjects of the rules under which the gamified environment works, including the behavioural interventions (e.g. the decisions the subjects will have to make and their implications). The *experimental treatments* are the different behavioural interventions whose impacts are measured by the experiment. As highlighted above, the critical part of the experimental design is that each subject needs to be randomly assigned to a treatment. The *behavioural measures* are the decisions made by the subjects in response to the behavioural interventions (i.e. the decision variables to be observed and studied).

The non-deception requirement

As in other disciplines involving research with human subjects, BEEs need to respect certain ethical considerations. They can never cause negative implications for participants, researchers, sponsors, implementers, future researchers, potential beneficiaries of the research, or the public at large. BEEs also need to achieve an additional ethical consideration, not required in other fields such as clinical trials or psychological experiments: namely, *non-deception*. Deception occurs when experimenters convey false or intentionally misleading information to participants. The use of deception in economic experiments is essentially forbidden, and the discipline's distaste for deception is often the first thing participants are informed of in economics experiments (Carmo Farinha, 2015, pg. 133). This requirement of non-deception makes the design of BEEs more complex but enhances the validity of the results.

The requirement for variable economic incentives

Not all behavioural experiments are BEEs. Beyond the non-deception requirement discussed above, there is another strict requirement that an experiment needs to fulfil to be considered a BEE: the application of an *incentivised research methodology* with *variable economic incentives* depending on the decisions made by the subject. This is the central feature of a BEE.

It differs significantly from a non-incentivised methodology in that although non-incentivised methodologies to research human behaviour can apply economic incentives, the amount paid to participants does not depend on their actual behaviour during the experiment. Non-incentivised research is based on the reaction of the subjects to a hypothetical stimulus, with no real impact on the subject. For instance, when participating in a standard survey-based conjoint analysis to determine the optimal configuration of a cybersecurity protection and cyber insurance portfolio, a subject could be asked to order a series of different configurations of cybersecurity protection and cyber insurance according to her preferences, then questioned as to which of these configurations she would actually purchase. In this case, the subject's choices during the research have no real impact on her and should be considered not as actual but merely hypothetical behaviour. Moreover, since no real cybersecurity decision is actually made in non-incentivised research, her state of mind is quite different from that when actual cybersecurity decision-making is involved.

To cope with the issue of choosing hypothetical alternatives, incentivised research methodologies with variable economic incentives generate a system of rewards for the subjects to respond to the factors presented in the different treatments in a realistic way. Note though that the simple presence of a reward is not sufficient for a method to be considered incentivised. According to Smith and Smith (1991), three conditions are required for a monetary incentive to induce value:

- *Monotonicity*, which means that subjects will always prefer a larger incentive (and they cannot reach saturation). This condition is easily fulfilled if the reward is paid in actual currency. Monotonicity motivates subjects to make the decision that maximises their final reward.

- *Salience*, which means that subjects' rewards are not fixed and will depend on their behaviour during the experiment (for instance, the purchase of cyber insurance or not) and random events (such as experiencing a cyber attack).

- *Dominance*, which means that any change in the utility of subjects is mainly determined by a change in their rewards and any other influence can be ignored.

The seminal work of Holt and Laury (2002) and of subsequent researchers have confirmed the existence of significant differences between the results obtained from incentivised and non-incentivised methodologies. Even when the monetary value of variable economic incentives is small, the information obtained from BEEs is more reliable than that obtained from similar non-incentivised experiments (Hernández and Vila, 2014). A possible explanation for this is that incentives not only increase the attention paid by subjects during the experiment but also lead subjects to the mental state of actual decision-making. This difference is critical when analysing behavioural interventions aimed at improving cybersecurity behaviour.

Behavioural Economics Experiment example

The CYBECO Project included three BEEs to analyse the potential for behavioural interventions to foster cybersecurity behavioural change. We present the design of one of these experiments to illustrate the role played by incentives in a BEE. This design was deployed online with more than 2,000 research subjects in four different EU countries. The subjects consisted of employees of SMEs and microenterprises with responsibility for their companies' cybersecurity decisions.

The experiment studied the subjects' cybersecurity strategy decisions (i.e. their allocation of financial resources between purchasing cybersecurity products and cyber insurance) on behalf of a fictional company in an online controlled gamified environment. Subjects were presented with different versions of a prototype "toolbox", an online tool also developed as part of the CYBECO Project for assessing a company's cybersecurity risk and providing advice on the optimal allocation of financial resources between cybersecurity products and cyber insurance.

The BEE allowed us to observe two behavioural measures: First, the subjects' cybersecurity strategy decisions, in response to the cybersecurity risk assessment and recommendations provided by the toolbox. Second, whether changes to the design of the toolbox's display page could influence the subjects' cybersecurity strategy decisions.

Running the experiment

To summarise how the experiment was run, each subject was assigned the role of IT head of an SME with responsibility for cybersecurity strategy decisions and was given an annual budget in Virtual Currency (VC) to spend. The subject was then taken through a cybersecurity risk assessment for the company using the toolbox. The toolbox also provided the subject with five cybersecurity strategy options for the company, ranked from best to worst, based on the company's cybersecurity risk assessment. (Note: The cybersecurity strategy options recommended by the toolbox were based on the model presented in Chapter 4.) Figure 3.4 shows a sample display screen, which presents the recommended cybersecurity strategy options as expected values framed as losses and includes a recommendation message.

The subject was then asked to choose their cybersecurity strategy, allocating their budget between the following options: (i) a choice of two different cybersecurity products (*Simple Security Measures* and *Advanced Security Measures*), which would reduce the company's probability of experiencing a cyber attack and (ii) a choice of two different cyber insurance policies (*Basic Insurance* and *Premium Insurance*), which would pay out if the company experienced an attack. They also had the option of taking out *No Insurance*. Figure 3.5 shows a sample display screen, which presents the cybersecurity products and cyber insurance options along with the price of each in VC.

Once the subject made their choice, the computer program randomly determined whether the company experienced a cyber attack. The subject was then shown a display screen calculating their total payout in VC. The payout was based on the choices they made in the experiment and whether the company had been attacked: more specifically, the budget remaining after purchasing cybersecurity products and/or cyber insurance, minus the losses generated by the cyber attack (should it have taken place), plus the coverage paid out by the insurance company (if cyber insurance was purchased). At the end of the experiment, the subject could exchange the VC into real euros. Figure 3.6 provides a sample display screen, which shows the payout for a subject who took out a cyber insurance policy and experienced a cyber attack.

As part of the experiment, different subjects were shown slight variations of the toolbox display screen. These differences included whether the top five cybersecurity strategy options shown on the display screen should: (i) be presented in terms of gains or losses (i.e. in terms of the financial benefits of purchasing cybersecurity products and cyber insurance or the potential financial harm due to not purchasing them); (ii) be presented in terms of the expected values of the gains or losses in the event of a cyber attack (computed from the probability of experiencing an attack); and (iii) include a message indicating that the first cybersecurity strategy option is recommended by cybersecurity experts along with a direct link for the subject to purchase it.

Note that the experiment (i) is non-deceptive and (ii) provides a variable economic incentive that is monotonic, salient, and dominant, meeting both of the requirements of BEEs. Also note that as part of the non-deception condition, it is not permitted to cheat the subjects in any way, in particular for the computation of their final payment. For instance, increasing the probability of suffering a cyber attack during the simulation to reduce the total payment to the subject would be an unethical practice that is not allowed in BEEs.

Findings

The experiment found that most of the subjects chose either the first or the second cybersecurity strategy options recommended by the toolbox. Almost 45% of subjects chose the first option, which was a cybersecurity strategy made up of Advanced Security Measures and Basic Insurance, while 33% chose the second option, a combination of Advanced Security Measures and Premium Insurance.

The experiment also found that the design of the toolbox's display screen had a significant impact on subjects' cybersecurity strategy decisions. The inclusion of a message on the display screen indicating that the first cybersecurity strategy option was recommended by cybersecurity experts along with a direct link for the subject to purchase the product increased subjects' propensity to choose the first option. Presenting the display screen in terms of expected values in the event of a cyber attack also increased subjects' inclination to choose the first option. This effect was particularly strong when the expected values were presented in terms of expected losses. This has important implications in terms of behavioural nudges that can be used to improve cybersecurity behaviour.

3.6 The benefits of combining psychological and behavioural economics approaches

In this final section we review some of the key lessons learned from a combination of psychological and behavioural economics models and consider the implications for both improving cybersecurity behaviour and the uptake of cyber insurance.

3.6.1 Cybersecurity compliance and cyber insurance uptake

Recent indicators suggest that research on cybersecurity behaviours and cyber insurance uptake would benefit from a combination of psychological and behavioural economics approaches. There is evidence that using PMT as an underlying behavioural model, coupled with constructs drawn from behavioural economics, can be beneficial in understanding organisational behaviour. A study by Bulgurcu et al. (2010) assessed the rationality-based factors associated with information security policy compliance, but also drew upon the TPB in evaluating the role of normative beliefs and individual self-efficacy in assessing the costs and benefits of compliance. Self-efficacy was indeed an important predictor, but they

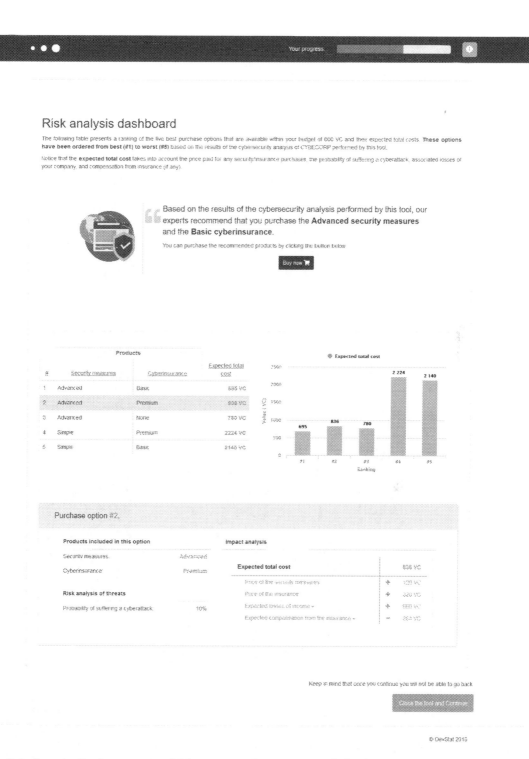

Figure 3.4: Sample display screen, which presents the recommended cybersecurity strategy options as expected values framed as losses and includes a recommendation message

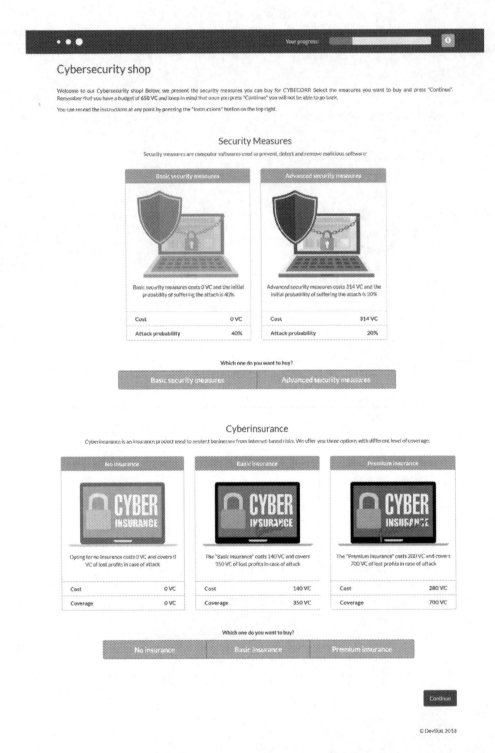

Figure 3.5: Sample display screen, which presents the cybersecurity products and cyber insurance options along with the price of each in VC

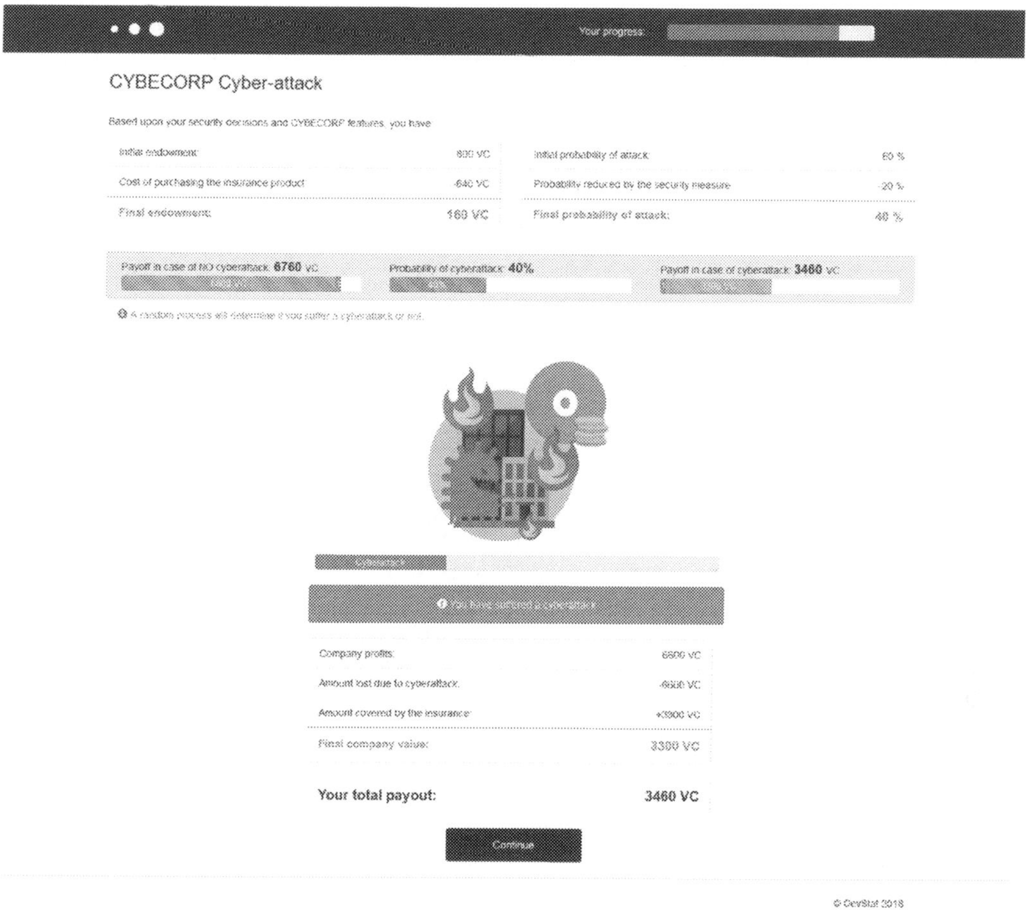

Figure 3.6: Sample display screen, which shows the payout for a subject who took out a cyber insurance policy and experienced a cyber attack

also found three broad classes of beliefs that influenced compliance: the benefits of compliance, the costs of compliance, and the costs of non-compliance. The authors argued that the *motivational factors* associated with information security policy compliance are important and that employees' cost-benefit assessments can be shaped by appropriate security training and awareness programmes.

Other psychological and economic constructs and models are also useful. A recent organisational paper on cybersecurity policy compliance draws upon both the psychological contract and rational choice literature in predicting behaviour (Han et al., 2017). The former is a large psychological literature describing the kinds of unwritten expectations employees have of the workplace and in this sense draws on the "shadow security" approach, in that the actual expectations of staff within the workplace are socially constructed. Their work is also interesting in that, yet again, the perceived *benefits*, rather than the *costs*, are what drive cybersecurity compliance behaviour.

A similar approach is adopted by Li et al. (2018), again using Rational Choice Theory to explore the underlying cost-benefit analysis of adopting good cybersecurity behaviours in the workplace, but this time adding a "procedural justice" construct as a further predictor of the extent to which employees are likely to adhere to cybersecurity policies. They also add an individual characteristic, "low self-control," in their study. This last component is interesting, as psychometric scales measuring a range of personality indicators are increasingly being adopted as a means to assess individual factors in poor compliance behaviour. In our own approaches, again adopting a combined psychology and behavioural economics ethos, we have sometimes used established measures of impulsivity and risk aversion (such as those of Jeske et al. (2016) and van Bavel et al. (2019)).

Such approaches are gaining ground, with greater prominence given to both psychological and behavioural economics models in EU guidance and policymaking (ENISA, 2019). Some authors, including de Bruijn and Janssen (2017), have argued that policymaking in cybersecurity is a "sea of paradoxes," in part because of the highly complex ecosystem that exists, which we discussed more fully in Chapter 2. They note that these paradoxes complicate the communication and framing of cybersecurity and argue that there is a greater need for theoretical framing of research. Using Prospect Theory as a backdrop, they call for evidence-based message framing strategies, including recognition of the need to avoid excessive use of fear appeals (as described in Section 3.4.4 on the Extended Parallel Process Model) and the need for messages to be framed in ways that are personally relevant to the recipient. This move away from fear appeals as a cornerstone of behaviour change towards a focus on building up appropriate competencies to manage threats at every level of the organisation is an important trend in cybersecurity research and is relevant to cyber insurance, where the ability to respond to threats appropriately can help to improve the resilience of an organisation. However, improving cybersecurity competence requires that employees at all levels take some responsibility for their actions and engage with training and awareness initiatives, even though they are busy with other tasks. For this reason, we end this chapter with a final reflection on incentives for behaviour change in this space.

3.6.2   Understanding incentives

Perhaps one of the most important issues to consider in future work concerns the way that incentives might influence cybersecurity behaviour. In BEEs, incentives are critical in determining the amount of cognitive effort participants put into their decision-making and psychological models such as PMT explicitly recognise the importance of rewards as a key behavioural influencer. In short, economic and psychological models and BEEs have demonstrated a key role for incentives. Yet an organisational challenge for cybersecurity, noted earlier, is that good cybersecurity behaviour is often disincentivised in the workplace since it is seen as slowing down productivity. It is thus rarely recompensed with an explicit reward (e.g. in terms of a promotion, bonus, or staff recognition). As noted above, this does not sit well with the desire to increase cybersecurity competency throughout an organisation. How can we incentivise better cybersecurity behaviour?

Can cyber insurance be used as an incentive? Or will it work as a disincentive, since liability for the cost of breaches lies elsewhere? We see these two arguments in the research literature. On the one hand, there is an argument that cyber insurance uptake drives an overall improvement in an organisation's cybersecurity posture, not least because the underwriters will only accept clients that have specific mechanisms in place to reduce risk (Khalili et al., 2018). It is possible to incentivise increases in an organisation's security posture through reduced premiums (and/or by requiring minimum controls before insurance will even be offered as a form of pre-screening) (Bailey, 2014; Khalili et al., 2018). Alternatively, or additionally, there is the possibility of imposing deductibles so that the insured organisation suffers some loss in the event of an incident (Young et al., 2016). There is also no doubt that insurers need to implement methods to allow continuous monitoring of cybersecurity practices (Young et al., 2016), which a number are already working on.

On the other hand, we are not seeing a great deal of this careful screening in practice and there is talk of a "race to the bottom" (selling cyber insurance policies cheaply and without onerous conditions) to ensure that the cyber insurance market takes off. Romanosky et al. (2019) have found that insurance providers "guess" the premiums for coverage, due to their lack of experience in the area (and also due to difficulty predicting intangible consequences such as loss of brand value) (Young et al., 2016). Some use reports that are significantly out of date to extrapolate the likelihood of a company being attacked. Additionally, many insurers follow a flat rate policy, where insured companies pay the same monthly premium regardless of personal circumstances: Only 31% of insurers in the study by Romanosky et al. (2019) used information about the clients' cybersecurity posture in the premium calculation process. Even those insurers who did apply factors based on a company's cybersecurity behaviours (e.g. measures such as privacy controls, network security controls, content liability controls, laptop and mobile device security policies, and incident report plans) used very broad categories, and the ratings for the behaviours were vague (e.g. average, above average, and below average).

Poor auditing of cybersecurity behaviour can increase the concern that the introduction of cyber insurance will lead to some kind of moral hazard (i.e. that it will generate more complacent behaviour as a result of offsetting risk). These concerns derive from the existence of moral hazard in relation to other types of insurance, such as car insurance (Cummins and Tennyson, 1996; Tolvanen, 2015). In behavioural economics terms, moral hazard would imply that cyber insurance could disincentivise careful System 2 thinking, as employees might consider the risks of poor cybersecurity choices to be lower when the reputational and financial costs are handled by the insurer. Yet we are not seeing this moral hazard argument play out in the available cyber insurance research literature. Indeed, in our own experimental studies, using the methodology described above, we found that cyber insurance does not appear to generate moral hazard. That is, purchasing a cyber insurance policy does not seem to increase risky behaviour or risky decision-making. In fact, we found evidence to suggest the contrary; that advantageous selection may play a role in cyber insurance uptake (i.e. individuals with insurance are risk averse and seek to reduce risk across all domains of their decision-making and behaviour). These findings suggest that concerns around moral hazard in the cybersecurity domain may be unfounded and that cyber insurance can play an important role in incentivising improved organisational posture.

3.7 Conclusions

In this chapter we have shown that psychological models of behaviour change can sit well alongside behavioural economics approaches in understanding the human factors at play in cybersecurity. We have also shown how BEEs can be designed to explore the role of individual factors in improving or impairing cybersecurity behaviour. Such issues are important when we seek to understand perception of risk and cyber insurance behaviours at the employee level as opposed to the organisational level. For individual employees, compliance with their organisational information security policies may be too onerous and as a consequence, they may display a dysfunctional response to cybersecurity risk—burying their head in the sand and leaving cybersecurity actions in the hands of others. Improving employee self-efficacy and driving up the cybersecurity competencies of individuals across the organisation is an important step and cyber insurance may be able to help rather than hinder this process, providing that employees are sufficiently well incentivised to change behaviour.

4

Risk Management Models for Cyber Insurance

Aitor Couce Vieira, David Ríos Insua
ICMAT

Caroline Baylon
AXA

Sebastain Awondo
University of Alabama

CONTENTS

In this chapter we develop a series of models to assist organisations and insurance companies with their decisions involving risk management in cybersecurity, making use of adversarial risk analysis and Multi-Agent Influence Diagrams to do so. Notably, we present a key model to support organisations in their Cybersecurity Risk Management decisions, referred to as "the CSRM model". The model helps organisations determine their optimal cybersecurity resource allocation, which is the best way for them to allocate their budget between spending on security controls and on cyber insurance. To make it simple for companies to use, we have also made available a prototype "toolbox" that provides an online interface for the model. In addition, we develop a series of auxiliary models to assist insurance companies. This includes a model to aid insurers with the design of cyber insurance products, helping them optimise price and coverage. It also enables the dynamic pricing of products. We present models for dealing with cyber insurance fraud as well, both a model for whether to issue a cyber insurance policy given the possibility of fraud and one for identifying fraudulent cyber insurance claims. Lastly, we put forth a model for making cyber reinsurance decisions, which makes it possible to better understand accumulation risk.

4.1   Introduction

This chapter sets forth solutions to some of the challenges for both organisations and insurance companies regarding cybersecurity and cyber insurance that were presented before. In Chapters 2 and 3, we discussed some of the difficulties that organisations face when making cybersecurity decisions, including those involving the purchase of cyber insurance. One factor is that organisations often have an inadequate understanding of the cybersecurity risks. Budgetary constraints, with a lack of sufficient investment in cybersecurity, are another key issue.

To address these problems, in this chapter we develop a new model to assist organisations with their Cybersecurity Risk Management (CSRM) decisions, which we refer to as "the CSRM model". This model helps organisations determine their optimal cybersecurity resource allocation, or the best way for them to allocate their budget between investing in security controls to defend against cybersecurity threats and in cyber insurance products to provide compensation in the event of an attack. To make it simple for companies to use the CSRM model, as part of the CYBECO Project we also developed a prototype "toolbox" that provides an online interface for the model. The toolbox is available at `https://toolbox.cybeco.eu/`.

In this chapter we also develop a series of additional models to assist insurance companies with their risk management problems in cybersecurity. We described some of the challenges that insurers face in Chapter 1. Notably, the difficulty in accurately assessing cybersecurity risk, making it difficult to develop and price cyber insurance products. We therefore propose a model to assist insurers with the design of cyber insurance products, helping them to optimise price and coverage. The model can be extended to engage in market segmentation and even allows for dynamic pricing of these products, i.e. for premiums to automatically adjust in response to an increase or decrease in an insured company's cybersecurity risk.

The other models that we have developed to assist insurers include those to deal with fraud. This consists of a model for assessing whether or not to issue a cyber insurance policy given the possibility of fraud and a model for identifying fraudulent cyber insurance claims. We present a model for determining the level of cyber reinsurance needed as well. This also makes it possible to better understand accumulation risk, or the risk that a single claim spreads to multiple lines of business, which is a major challenge for insurers.

Overview

The models developed draw heavily on Adversarial Risk Analysis (ARA), a novel subfield of decision analysis which helps a decision maker maximise his expected utility while taking into account the strategies and uncertainties of his adversaries. ARA draws extensively on the use of Multi-Agent Influence Diagrams (MAIDs), which are visual displays of decision-making situations. The next section, Section 4.2, provides a basic introduction to ARA and MAIDs.

In the subsequent sections, we describe each of the models that we have developed in detail. Section 4.3 presents the CSRM model for cybersecurity resource allocation and cyber insurance product selection. Section 4.4 presents the model for cyber insurance product design as well as ways to extend the model. Section 4.5 presents the models for the issuing of cyber insurance policies and for fraud detection. Section 4.6 presents the model for cyber reinsurance decisions.

When describing the models, we first present the relevant use cases, in order to ensure that they are grounded in practical experience. We employed expertise from both the cybersecurity and insurance sectors in creating the use cases, and they are based on the value chain of a company and its assets. Next, we discuss the current approaches being used to address the problems described in the use cases and the drawbacks with these approaches. We then formulate the models we have developed to overcome the use case problems, describing how we conducted the ARA and presenting the associated MAIDs. Finally, we show how to solve the models in mathematical terms.

This chapter remains at the conceptual level, describing generic modelling and computational issues. The models that we present here can serve as basic templates that can be expanded, reduced, or otherwise modified, according to the need. An Appendix provides a simple numerical illustration. For further detail, Chapter 5 sets forth a full case study for the CSRM model.

4.2   Methodology

The models for managing cybersecurity risk that we develop in this chapter are based on the use of adversarial risk analysis, which relies extensively on the use of Multi-Agent Influence Diagrams. This section provides an overview of these methodologies.

Adversarial Risk Analysis

Adversarial Risk Analysis, a novel subfield of decision analysis that also draws on statistical risk analysis and game theory, enables a decision maker to best allocate his resources to defend against several adversarial decision makers, taking their strategies as well as uncertainties into account (Banks et al., 2015).

In ARA, we seek to maximise the expected utility of a decision maker, whom we refer to as the "Defender", so that she makes the optimal choice between a set of decision alternatives. We take into account her expected utilities given the conditional probabilities of various threats and impacts occurring. This includes calculating the conditional probabilities of the Defender being attacked by various other decision makers, who are adversaries of the Defender and whom we refer to as the "Attackers". To do so, we use simulation to determine the expected utilities of the Attackers when it comes to attacking the Defender. This involves considering their decision alternatives. Since we have limited information about the Attackers, we also incorporate uncertainty regarding their strategies, beliefs, and preferences by making use of subjective probability distributions. We then use optimisation to determine the Defender's decision strategy that maximises her expected utility in the given adversarial situation. The types of problems that we solve with ARA can be referred to as simulation-optimisation problems, because we use simulation to forecast the Attackers' behaviour and optimisation to determine the best strategy for the Defender given the circumstances.

ARA draws on statistical risk analysis to model the probability of uncertain outcomes as well as on game theory to model reasoning about the strategies of the adversaries. However, it diverges from risk analysis in that it assumes that the adversaries are intelligent and think strategically, and it departs from standard game theory because it does not make the assumption that the decision makers have common knowledge of each other's preferences and beliefs. It also does not seek an equilibrium among the decision makers, since it solves from the perspective of the Defender rather than solving for all decision makers' problems jointly.

The use of ARA is important for developing the cybersecurity risk management models presented in this chapter because it allows us to model the specific threats that are targeting an organisation. The sheer number of cybersecurity threats mean that organisations have to prioritise strategically when it comes to defending against them. This is especially important given the budget pressures that they face.

Multi-Agent Influence Diagrams

We represent the problems and models graphically with the aid of Influence Diagrams (ID), which are visual representations of decision situations. They are often considered an extension of Bayesian networks (French and Insua, 2000). We use the terminology used in Banks et al. (2015): The general term Multi-Agent Influence Diagram (MAID) can be used to describe decision situations that involve more than one agent (i.e. decision maker). We can also use more specific terms, such as Bi-Agent Influence Diagram (BAID) to describe decision situations involving two agents and Tri-Agent Influence Diagrams (TAID) for those involving three agents.

The diagrams involve four types of nodes:

- *Uncertainty nodes*, which correspond to each uncertainty that needs to be modelled. Each has an associated conditional probability distribution. They are sometimes also called chance nodes. Uncertainty nodes are represented as an *oval* in these diagrams.

- *Deterministic nodes*, which are a subtype of uncertainty nodes. They are a special type of uncertainty where the outcome is known when the outcomes of certain other uncertainties are known (i.e. the values of the nodes at their tails, that is, the nodes that immediately precede them). Deterministic nodes are represented as a *double oval* in these diagrams.

- *Decision nodes*, which correspond to each decision that the agent has to make from a set of alternatives. Decision nodes are represented as a *rectangle* in these diagrams.

- *Value nodes*, which model the agent's preferences, as determined by their utility function. Value nodes are represented as a *hexagon* in these diagrams.

The diagrams also involve two types of arrows:

- *Dotted arrows*, which point to decision nodes. The dotted arrows indicate that decisions are made knowing the values of all the nodes at their tails beforehand.

- *Solid arrows*, which point to uncertainty nodes or value nodes. They can also point to deterministic nodes. The solid arrows mean that the corresponding events or consequences are influenced by the values of the nodes at their tails.

The nodes that are relevant to just one agent's decisions are in solid colours (with each agent represented by a different colour, e.g. white, light grey, or dark grey). Nodes that are relevant to several agents' decisions are striped.

The use of Multi-Agent Influence Diagrams are important when developing our models for cybersecurity risk management in order to help us visualise complex decision situations. For this reason, they are a valuable starting point when conducting an ARA.

4.3 The CSRM model: Cybersecurity resource allocation, including cyber insurance product selection

In this section we develop a CSRM model to help an organisation determine how to best allocate its cybersecurity resources. This includes providing advice on the selection of a cyber insurance product. Specifically, the model calculates the optimal way for the organisation to apportion its budget between spending on security controls and on cyber insurance.

The model applies to companies of all sizes. We use an SME in the use case below given that Chapters 2 and 3 found that the cybersecurity challenge may be particularly acute for SMEs for a number of reasons: For one, they are a primary target for attacks. Compounding the problem, SMEs are particularly vulnerable, given that they are even less likely than larger companies to have the necessary expertise to understand the cybersecurity risks. They also have fewer resources available to invest in cybersecurity. SMEs must therefore make a number of trade-offs when it comes to cybersecurity investment, making it all the more important for SMEs to optimise their cybersecurity resource allocation.

4.3.1 Use case

A decision maker within an SME must decide which cyber insurance product to buy. The decision maker is considering three different cyber insurance products, which provide increasing levels of coverage:

1. *Cyber Insurance Product 1* covers the risk of internal data loss, including the loss of competitive advantage following the theft of a company's customer database, intellectual property, and business intelligence.

2. *Cyber Insurance Product 2* covers the risk covered by Cyber Insurance Product 1 as well as the risk of brand damage following a data breach, including but not limited to the loss of sensitive customer data, investor divestment, or a decline in share price and in the company's investment rating due to negative media coverage.

3. *Cyber Insurance Product 3* covers the risks covered by Cyber Insurance Product 2 as well as the risk of failure to deliver products and services, that is, non-fulfilment of service and contractual agreements with respect to third parties and the ensuing impact on the assets and business continuity of these third parties.[1]

These cyber insurance products might be offered by a single insurance company that has a range of products or they might be offered by different insurance companies. They could be sold directly by the insurance company or companies or by an insurance broker who acts as a middleman.

In deciding which cyber insurance product to buy, the decision maker considers several factors, including the criticality of the SME's assets. For example, if compromising a certain asset in the SME's value chain could threaten its solvency, then the decision maker might consider that it is essential to insure against this risk. As described in Chapter 2, the decision maker also takes the company's regulatory and other legal obligations into account, which may also prompt him to purchase cyber insurance to cover risks associated with these obligations.

---

[1] Note that this book focuses on standalone cyber insurance policies. However, cyber insurance coverage can and does exist under other lines of business.

4.3.2 Current approaches

A variety of methods have been developed that can be used to support decision-making regarding cybersecurity resource allocation, including the purchase of cyber insurance. As mentioned in Chapter 1, there are a large number of CSRM frameworks. These include ISO 27005, an international risk management standard; CORAS, a security risk analysis method; and MAGERIT, a risk analysis and management framework. However, these approaches tend to be based on risk matrices, which have a number of shortcomings also covered in Chapter 1. Notably, they systematically assign the same rating to threats that are very different qualitatively, potentially leading to the sub-optimal allocation of cybersecurity resources.

Other approaches make use of optimisation and game theory models, and combinations of the two, as reviewed in Fielder et al. (2016). Schilling and Werners (2016) and Khouzani et al. (2019) describe combinatorial optimisation approaches for the optimal selection of IT security safeguards, but they do not take risk or adversarial aspects into account. Cavusoglu et al. (2008) and Rao et al. (2016) use game theory approaches for cybersecurity resource allocation; however, they make common knowledge assumptions which might not be realistic in the cybersecurity context.

4.3.3 Model formulation

To overcome the issues with the existing approaches described above, we formulate a new CSRM model to assist organisations with their cybersecurity resource allocation decisions, including the purchase of cyber insurance. In doing so, we make use of Adversarial Risk Analysis to take adversarial aspects into account. This relies heavily on Multi-Agent Influence Diagrams, a key component of ARA. Further details may be seen in Rios Insua et al. (2019) and Couce-Vieira et al. (2020b).

We first create a model of an organisation's CSRM decision-making process, presented as a Tri-Agent Influence Diagram in Figure 4.1. The three agents in this diagram are the SME deciding on its cybersecurity resource allocation (designated as the *Defender*) and two types of attackers (designated as *Attacker 1* and *Attacker 2*). Attackers might range from cybercriminals to competitors to nation states, while their motives for attacking the Defender could include anything from financial gain (e.g. stealing customer data, including credit card information) to obtaining competitive advantage (e.g. stealing business information from a competitor) to causing physical harm through cyber-physical means. To distinguish between the agents, nodes involving the Defender are white, nodes involving Attacker 1 are light grey, and nodes involving Attacker 2 are dark grey; striped nodes are relevant to several agents.

The TAID further expands upon the schematic view of an organisation's CSRM process presented in Chapter 1 (Section 1.2). The *organisation profile and features* (designated as ft) describes the organisation in terms of its profile, assets, and other features and is modelled as a deterministic node, which we thus represent as a double oval.

We next identify the threats that may impact the organisation. These include *targeted cyber threats*, which are cyber attacks that are intentionally directed at the organisation by Attacker 1 (designated as tc_1) or by Attacker 2 (designated as tc_2). They are modelled as decision nodes (since they involve decisions made by the attackers) and thus represented as rectangles.

As suggested by the Information Security Forum (ISF), we also include the following unintentional threats: *non-targeted cyber threats* (designated as ntc), which are cyber attacks that are not directly targeted at the organisation (e.g. a malware campaign that happens to hit the organisation); *accidental threats* (designated as a), which are cyber incidents caused

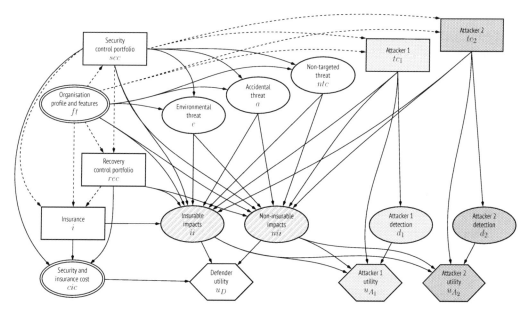

Figure 4.1: TAID of the CSRM problem for an organisation. White nodes correspond to the Defender, light grey modes correspond to Attacker 1, dark grey nodes correspond to Attacker 2, and striped nodes affect several agents.

without malicious intent (e.g. data loss caused by an employee accidentally deleting files); and *environmental threats* (designated as e), which involve floods, earthquakes, and other natural disasters that may impact the organisation's IT equipment. These are all modelled as uncertainty nodes and thus represented as ovals.

Having identified the threats, we determine the impacts that are relevant for the organisation. These include both *insurable impacts* (designated as ii) that are partly covered by insurance and *non-insurable impacts* (designated as nii) that are not covered by insurance. They are modelled as uncertainty nodes so are represented as ovals, and are striped because they are relevant to several agents' decisions.

The organisation can choose from a number of tools to mitigate the likelihood and/or impacts of the threats. These include two different types of security products: the *security control portfolio* (designated as sec), which are security products and other measures to protect and defend against cyber attacks, and the *recovery control portfolio* (designated as rec), which are security products and other measures to recover from attacks, including those to restore systems and information. They also include *insurance* (designated as i), which are insurance contracts used for risk transfer purposes. These instruments are all modelled as decision nodes (since they involve decisions made by the organisation) and thus depicted as rectangles.

These tools may need to take certain constraints into account, including (i) legal and regulatory, such as the EU's General Data Protection Regulation (GDPR), (ii) financial, such as the organisation's cybersecurity budget, (iii) compliance, such as cybersecurity measures mandated by national regulators, or (iv) technical, such as not being able to adopt some cybersecurity products on the market due to interoperability issues with the IT products that the organisation already has. They will also entail *security and insurance costs* (designated as cic), which is a deterministic node and hence denoted as a double oval.

Having identified all of the elements that are relevant for the Defender, we can now build the Defender's preference model, which is determined by her utility function. We therefore model the *Defender's utility* (designated as u_D) as a value node and thus depict it as a hexagon. This utility function covers risk aversion aspects as well.

Next, we turn to the remaining elements that are relevant for the attackers. To take Attacker 1 as an example, a relevant factor is the *detection of Attacker 1* (designated as d_1), which is whether or not the organisation is able to identify Attacker 1 as being responsible for the attack. This is modelled as an uncertainty node and hence depicted as an oval. Sophisticated attackers engaging in targeted attacks tend to perpetuate these attacks with information they have invested time and effort in uncovering during a reconnaissance phase, including the organisation profile and features ft and its security control portfolio *sec*.

Having identified all of the elements that are relevant for the attackers, we can build the preference model for an attacker, which is also based on his utility function. We therefore model *Attacker 1's utility* (designated as u_{A_1}) as a value node and accordingly depict it as a hexagon. The process is analogous for Attacker 2.

4.3.4   Model solution

To determine an organisation's optimal resource allocation for what we call its "cybersecurity portfolio," which is a combination of security products (its security control portfolio and its recovery control portfolio) and insurance products (notably cyber insurance), we follow these steps:

1. We examine the CSRM problem from the perspective of the Defender, including creating the Defender's Influence Diagram. We build the Defender's preference model (that is, her utility function) regarding her optimal resource allocation, taking into account the Defender's decision alternatives between different types of security products (security control portfolio and recovery control portfolio) and insurance options, as well as the conditional probabilities of various threats or impacts occurring.

2. In order to determine the conditional probabilities of attacker threats occurring—i.e. the probability of the Defender being attacked by either Attacker 1, Attacker 2, or both—we need to examine the CSRM problem from the perspective of the Attackers; that is, the Attackers' strategic thinking when it comes to attacking the Defender. We create the Attackers' Influence Diagrams and build the Attackers' preference models (their utilities) regarding whether to carry out certain targeted attacks. We take into account their uncertainties regarding such factors as whether they will be detected and the attack impacts as well as their decision alternatives in terms of the various types of targeted attacks to carry out. We then use simulation in order to forecast the Attacker's likelihood of attacking the Defender.

3. Finally, we use optimisation in order to maximise the Defender's expected utility given the Attackers' strategic thinking about attacking the Defender. We take into account the Defender's probability of being attacked by either or both of the Attackers as well as each feasible cybersecurity portfolio. This yields the cybersecurity portfolio that gives the Defender the maximum expected utility, revealing her optimal choice of a cybersecurity portfolio to defend against the Attackers.

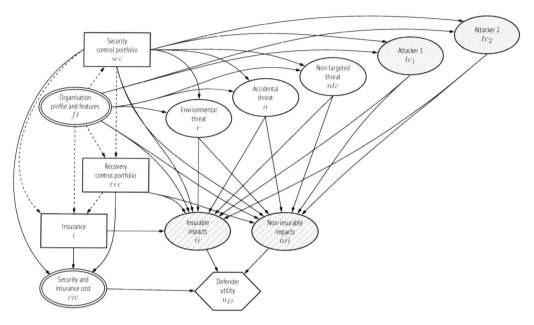

Figure 4.2: ID of the CSRM problem for the Defender

The Defender's problem

We begin by drawing out the Defender's part of the CSRM problem presented in Figure 4.1. We represent this as an Influence Diagram in Figure 4.2.

Next, we develop the relevant mathematical formulations to describe each node:

- At the *uncertainty nodes* [oval], we define the conditional probability distributions of each threat or impact occurring for each node. (Each node represents either a threat or an impact.) For instance, for the accidental threat a node, the accidental threats a that could affect the organisation depend on the security control portfolio sec that has been implemented by the organisation and on the organisation profile and features ft. We therefore use the distribution $p_D(a|sec, ft)$ to represent the Defender's probability that accidental threat a occurs, given the implemented security control portfolio sec and the organisation profile and features ft. The threats used for the model could be chosen from a threat catalogue (such as that published by the ISF) and the distributions from a relevant parametrized distribution, which are assumed conditionally independent given sec and ft, as illustrated in Chapter 5.

- At the *deterministic nodes* [double oval], we express the values of each node as functions of the values of the nodes at their tails (i.e. the nodes that feed directly into them). For example, for the security and insurance costs cic node, the security and insurance costs cic depend on the costs of the security control portfolio sec, of the recovery control portfolio rec, and of the insurance product i. We therefore represent the security and insurance costs cic by the function $cic = g(sec, rec, i)$, for a certain function g.

- At the *decision nodes* [rectangle], we define the set of security or insurance products that the Defender can choose from for each node. (Each node represents either different types of security products, i.e. either security control products or recovery control products, or else insurance products.) For example, for the recovery control portfolio rec node we define the various recovery control portfolios rec (e.g. different types of recovery control products) that the organisation can purchase. These are chosen knowing the organisation's security control portfolio *sec* and its organisation profile and features *ft*, as well as any other relevant constraints. We can therefore define the recovery control portfolios *rec* that satisfy the relevant constraints by *rec(sec, ft)*.

- At the *value node* [hexagon], we define the preference model for the Defender, which is determined by her utility function. Her utility depends on the security and insurance costs *cic*, the insurable impacts *ii*, and the non-insurable impacts *nii*. We can therefore express her preference model as the utility function $u_D(cic, ii, nii)$.

    The above assessments are standard and may be based on data and/or expert judgement (Clemen and Reilly, 2013). The exception to this is at the uncertainty nodes for the Attackers, when determining the Defender's probability that it will experience an attack tc_1 carried out by Attacker 1 or an attack tc_2 carried out by Attacker 2. This is expressed as $p_D(tc_i|sec, ft), i = 1, 2$ given security control portfolio *sec* and organisation profile and features *ft*. Here, we must take into account elements of the Attackers' strategic thinking.

The Attacker's strategic thinking

To assess the Defender's probability of being attacked by Attacker 1, Attacker 2, or both, we must consider the Attackers' components of the CSRM problem presented in Figure 4.1. Using Attacker 1 as an example, we present an Influence Diagram for Attacker 1 in Figure 4.3. The diagram is based on the belief that the Attacker has means to detect security control portfolio *sec* and organisation profile and features *ft*, which for sophisticated attackers engaging in targeted attacks is a reasonable assumption. The Influence Diagram is analogous for Attacker 2.

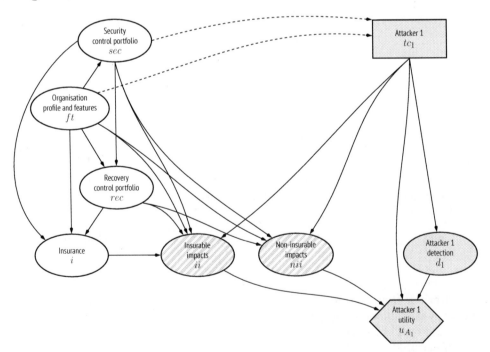

Figure 4.3: ID of the CSRM problem for Attacker 1

Since we tend to have limited information about the Attacker, we have uncertainty about his beliefs and preferences. We therefore develop the following mathematical formulations to describe them:

- At the *uncertainty nodes* [oval], we define the conditional probability distributions for the nodes that are relevant. (This includes the nodes for whether or not the organisation detects an attack and for the impacts of attacks.) We use random probability models to assess our uncertainty about Attacker 1's beliefs, identifying the corresponding random conditional probability distributions for each node. For example, we are not able to ask Attacker 1 what he believes his probability of being detected is if he carries out a targeted attack tc_1, which we represent as $p_{A_1}(d_1|tc_1)$. We therefore model our uncertainty about the distribution through a random distribution $P_{A_1}(d_1|tc_1)$ (a distribution over the probabilities $p_{A_1}(d_1|tc_1)$).

- At the *decision node* [rectangle], we define the set of targeted attacks that the Attacker can choose from to wage against the Defender. These are chosen knowing the organisation's security control portfolio sec and its organisation profile and features ft, as well as any other relevant constraints. We can therefore define the targeted attacks tc_1 that satisfy the relevant constraints by $tc_1(sec, ft)$.

- At the *value node* [hexagon], we define the preference model for Attacker 1 based on his utility function. His utility depends on the impacts of the attack (both insurable impacts ii and non-insurable impacts nii), whether the Defender detects the attack d_1, and on the targeted cyber attack that he directs at the Defender tc_1. Again, as we are not able to ask Attacker 1 or otherwise obtain information to assess his utility function, we model our uncertainty about it with a random utility function $U_{A_1}(ii, nii, d_1, tc_1)$ (a distribution over the utilities $u_{A_1}(ii, nii, d_1, tc_1)$). This gives us his preference model.

Then, given a security control portfolio *sec* and organisation profile and features ft, we find the random expected utility (a distribution over the expected utilities) associated with Attacker 1 carrying out a targeted attack tc_1 through

$$\Psi_{A_1}(tc_1|sec, ft) = \int ... \int U_{A_1}(ii, nii, d_1, tc_1) \times$$
$$\times P_{A_1}(nii|tc_1, sec, rec, ft) P_{A_1}(ii|tc_1, sec, rec, i, ft) P_{A_1}(d|tc_1) \times$$
$$\times P_{A_1}(rec|ft, sec) P_{A_1}(i|sec, rec, ft) \, dnii \, dii \, di \, dd_1 \, drec.$$

We next determine the random optimal attack (the distribution over the optimal attacks) maximising it

$$TC_1^*|sec, ft = \arg\max_{tc_1} \Psi_{A_1}(tc_1|sec, ft). \tag{4.1}$$

Finally, we estimate the required probability distribution through

$$p_D(tc_1|sec, ft) = Pr(TC^* = tc_1|sec, ft).$$

We can carry out this same process for Attacker 2.

Finally, we consider the probability that both Attackers might attack the Defender. Assuming conditional independence and given the portfolio implemented and the organisation profile and features ft, we calculate

$$p_D(tc_1, tc_2|sec, ft) = p_D(tc_1|sec, ft) p_D(tc_2|sec, ft).$$

Maximising the Defender's expected utility given the Attackers' strategic thinking

To determine the optimal resource allocation for the Defender, we use the Defender's probability of being attacked by each Attacker, expressed as $p_D(tc_i|sec, ft), i = 1, 2$ for a given security control portfolio sec and organisation profile and features ft.

For each feasible combination of security products (which includes both the security control portfolio sec and the recovery control portfolio rec) and insurance products i, we associate them with their expected utilities and find the portfolio with the Defender's maximum expected utility according to the following steps:

1. Remove the security and insurance cost cic deterministic node, computing the utilities

$$u_D\big(g(sec, rec, i), ii, nii\big).$$

2. Compute the portfolio expected utility, removing the uncertainty nodes

$$\psi(sec, rec, i|ft) = \int \cdots \int u_D\big(g(sec, rec, i), ii, nii\big) \times$$
$$\times\, p_D(nii|tc_1, tc_2, ntc, a, e, sec, rec, ft) p_D(ii|tc_1, tc_2, ntc, a, e, sec, rec, i, ft) \times$$
$$\times\, p_D(tc_1, tc_2|sec, ft) p_D(ntc|sec, ft) p_D(a|sec, ft) p_D(e|sec, ft)$$
$$\mathrm{d}nii\ \mathrm{d}ii\ \mathrm{d}cic\ \mathrm{d}tc_1\ \mathrm{d}tc_2\ \mathrm{d}ntc\ \mathrm{d}a\ \mathrm{d}e. \quad (4.2)$$

3. Remove the insurance i decision node, obtaining and taking note of the optimal cyber insurance product, given the security portfolio

$$i^*(sec, rec|ft) = \arg\max \psi_D(sec, rec, i|ft),$$

$$\psi_D(sec, rec|ft) = \psi_D(sec, rec, i^*(sec, rec|ft)|ft).$$

4. Remove the recovery control portfolio rec decision node, obtaining and taking note of the optimal recovery controls, given the optimal security controls

$$rec^*(sec|ft) = \arg\max\ \psi_D(sec, rec|ft), \quad \psi_D(sec|ft) = \psi_D(sec, rec^*(sec|ft)|ft).$$

5. Remove the security control portfolio sec decision node, obtaining and taking note of the optimal security controls, given the organisation profile and features ft

$$sec^*(ft) = \arg\max \psi_D(sec|ft).$$

The optimal resource allocation would then be

$$(sec^*(ft),\ rec^*(sec^*(ft)|ft),\ i^*(rec^*(ft), rec^*(sec^*(ft)|ft))).$$

As mentioned earlier, the model described here is a basic template. For example, it could be simplified by eliminating the rec node if it is not relevant. Or it could be made more complex by including more than two types of Attackers. It could also include different or additional types of data. For instance, health data is commanding particularly high prices on the dark web at present, with financial data not far behind. In Chapter 5, our case study illustrates how this can be done, providing detailed numerical examples.

4.4   Cyber insurance product design

We now turn to the models that we have developed to help insurance companies with their risk management problems in cybersecurity. In this section we propose a model to assist insurers with the design of cyber insurance products. The model makes it possible to optimise both price and coverage. The model can be adapted to enable market segmentation and also to allow for dynamic pricing of these products.

This model is designed to help with some of the challenges for insurers that are described in Chapter 1, notably the difficulty surrounding developing and pricing cyber insurance products. This stems in large part from the struggle to accurately assess cybersecurity risk. The use case below describes a situation in which an SME receives insufficient cyber insurance coverage, due to experiencing a cyber attack that has larger than anticipated impacts on third party companies that rely on its products or services.

4.4.1   Use case

During a large DDoS attack by a nation state-sponsored attacker, an SME that provides cloud-based services to several major companies is unable to do so. Since these companies rely on the SME's cloud-based services for a significant part of their operations, they are also unable to provide services for up to 12 hours. This third party business interruption entails a large financial impact that greatly surpasses the initial risk assessment estimation.

From the perspective of the insurance company, this could be addressed by applying a maximum cap for claims related to cyber incidents. However, from the perspective of the insured company, it is important that it conducts a correct assessment of the impact of a cyber incident on its entire value chain, including third party companies.

4.4.2   Current approaches

We now describe how cyber insurance products are designed at present. Information about the specific algorithms used by insurers in pricing cyber risk is confidential, but we can describe the process in general terms. Sarabia et al. (2007) provide a good overview of the design of insurance products more broadly, and Romanosky et al. (2019) give a detailed description of current practices in the US concerning cyber insurance product design. It is important to note that the decision process is frequently market-driven, in the sense that companies do not want to deviate too much from the prices that their competitors charge.

Cyber insurance product lines are currently limited, the main types of coverage being for cyber attacks, data leaks, forensic investigation costs, business interruption costs, and reputational damage, as described in Chapter 1. Some insurers specialise in specific types of cybersecurity coverage and market segments, in order to minimise the probability of major losses, develop expertise and reputation, and gain cyber insurance market share. For example, an insurer may choose to underwrite limited first party cyber breaches (e.g. hardware and software restoration or legal costs) for homogenous firms (e.g. financial, health, or technology) with a limited number of employees (e.g. less than 1,500) and vendors.

Key decisions for insurers underwriting cyber insurance products surround pricing and level of coverage, including when it comes to which impacts are insured. Some of the potential impacts of cyber attacks are mentioned in the various use cases in this chapter and in Chapter 5. The financial impacts of cyber incidents are described in Eling and Wirfs (2019). A review of business and organisational impacts in cybersecurity, beyond the traditional technical triad of confidentiality, integrity, and availability of information, may be seen in Couce-Vieira et al. (2020a).

The underwriting process

The cyber insurance underwriting process is still evolving and depends on the type of sensitive personal data (e.g. credit card numbers, social security numbers, or health information) and/or business-related data stored by the company, the type of liability to be transferred to the insurer, and the company's potential losses.

Preliminary information collection

In general, prior to drafting a contract, the insurer conducts a thorough risk assessment of the prospective client using in-house expertise or through a third party company. The first step involves collecting information that allows the insurer to assess the company's cyber risk profile. Insurers will request information on:

- the amount of resources dedicated to information security within the company, e.g. whether the company has a Chief Information Officer and if so, what their primary responsibilities entail;

- their information security procedures and to what extent they are enforced;

- the security measures in place, such as network segmentation, log monitoring, patch management, and encryption;

- the level of employee training on information security and whether there is an awareness campaign within the company, e.g. fake phishing attempts;

- their incident response plan and its testing frequency;

- their business interruption and recovery plan and its testing frequency;

- their procedures for third party vendor management, including the contractual obligations of vendors; and

- board-level oversight of cybersecurity policies and reported incidents.

The insurer then uses this information to estimate the probability of the company being breached by a cyber attack and the severity of the breach, given the security control and recovery control measures and procedures in place in the company. They also estimate the impacts on all stakeholders (not just on the company but also on the individuals whose data the company holds), based on specific risk measures.

The process itself

Broadly speaking, the cyber insurance underwriting process involves:

- modelling the frequency and severity of cyber losses,

- using the estimated models to simulate a joint probability distribution of breach frequency and severity,

- converting the breach frequency and severity distributions into monetary value,

- quantifying the risk of monetary losses with specific risk measures,

- estimating the cost of capital required to finance and administer the insurance policy, and

- setting gross premiums for the policy.

Modelling elements

We further describe the modelling aspects: As mentioned previously, cyber insurance underwriters start by modelling the frequency and severity of claims (aggregated claims). To do so, they make widespread use of the loss distribution approach (LDA) and copula-based models, as with other insurance business lines (Awondo, 2019; Eling and Jung, 2018). To use these models to simulate the joint probability distribution of breach frequency and severity, copulas are commonly used to model dependency structures of aggregated claims for each breach type across industries, due to their flexibility in modelling multivariate dependency. If there is sufficient data available, this can be done with a non-parametric estimation of the individual marginal distributions. However, when it comes to cybersecurity, data tends to be limited, so generally expert judgement methods are used instead, as in Couce-Vieira et al. (2020a).

Once the copula and individual marginal distributions are estimated, then the underwriters can simulate the multivariate distribution. The data is converted into monetary value and risk measures such as the value-at-risk (VaR), the conditional value-at-risk (cVaR), or the expected shortfall are used to quantify the risk involved (Rockafellar and Uryasev, 2002; Sklar, 1959). To model losses above a certain threshold, the Generalized Pareto Distribution (GPD) is typically used.

Setting the premium

The gross premium charged for a policy is the sum of the expected loss, also known as the *actuarially fair premium*. This is made up of a loading cost η for potentially catastrophic risk, which is usually a function of the tail risk or variance, and a loading cost for the capital required to administer the insurance policy. Insurers typically consider η as the cost of the investors' capital required for financing the policy. This estimate depends on how much money the insurer needs to set aside, and for how long, in the event of claims stemming from catastrophic losses. Several different approaches have been proposed for calculating η and the preferred choice is case-specific. A comprehensive estimate of η accounts for inflation, deductibles, policy limits, the cost of the investors' capital, and reinsurance. As mentioned, the final gross premium charged should be in line with the cost of similar policies sold by competitors. This ensures that the insurer can attract and sustain sufficient demand to significantly reduce the probability of excessive losses and be profitable.

Designing the policy and drafting the contract

To incite insured companies to minimise their risk and reduce moral hazard, insurers typically employ instruments such as co-insurance, deductibles, and caps on insured losses. Important components of the policy design include defining:

- exclusions,

- when the policy is triggered (often when a claim is made against the insured),

- when the insurer has to be notified, and

- which forensic, legal, public relations, and crisis management experts will be used following a breach.

With these components in place, the insurer proposes a policy to the client, which the client often further negotiates, and then issues a final version. Finally, the policy is activated and enforced when the client pays the premium and signs the contract.

4.4.3 Model formulation and solution

To solve these challenges, we now present the model we have developed for optimising price and coverage. We then show how it can be adapted to allow for market segmentation and dynamic pricing.

Optimising price and coverage

Price

We present an approach to pricing a cyber insurance product for a given client, determining the maximum price that the company would be willing to pay to include an insurance product in his optimal cybersecurity portfolio. Using the methods described in Section 4.3.4, we compute the expected utility u_D of the optimal portfolio x^*, which can be written concisely as

$$\arg \max_x \int u_D(x, p(i), \theta) p_D(\theta|x) d\theta. \tag{4.3}$$

We make explicit the dependence on the insurance pricing decision through a utility function which includes the price $p(i)$ of the insurance product; i.e. the solution will be $x^*(p(i))$. Here $x = (sec, rec, i)$ designates a generic portfolio, θ encompasses the relevant random variables ($nii, ii, tc_1, tc_2, ntc, a$, and e), and p_D is the distribution modelling the relevant uncertainties, including that of the company being attacked. The price $p(i)$ should be in a relevant range $[a, b]$, where a and b are, respectively, the minimum and maximum prices charged by competitors.

For a given insurance product i that an insurer is interested in selling to a company, we can determine the maximum price $\widehat{p}(i)$ for which the company would include such a product within his optimal cybersecurity portfolio x^*. Typically, the product arises within an analysis as described in Section 4.3.4 leading to the determination of the company's optimal cybersecurity portfolio, which we designate as the "reference portfolio", based on the initial reference price and coverage. We aim to find the maximum price $\widehat{p}(i)$ that the company would be willing to pay given his organisation profile and features, including his risk profile.

The maximum price $\widehat{p}(i)$ can be determined by searching in a grid of possible prices starting from the reference price and going all the way to the right extreme b (the highest price charged by his competitors). Since for the reference price we know that the optimal portfolio includes the insurance product, we first check whether for the next higher price in the grid the optimal portfolio still includes the insurace product; if so, we move further up the grid until it is not included or we reach b. We can then refine the search within the last interval identified. This gives us the maximum price that the company is willing to pay for an insurance product in his optimal cybersecurity portfolio.

Coverage

We also present an approach to determining the level of coverage of a cyber insurance product for a given client, based on the minimum coverage of impacts that the company would need in order to include the product in his optimal cybersecurity portfolio. We can use a similar approach to that used above. First we specify the dependence $p_D(\theta|x, c(i))$ on the coverage $c(i)$, so that we rewrite (4.3) as

$$\arg \max_x \int u_D(x, \theta) p_D(\theta|x, c(i)) d\theta.$$

Next we undertake a similar grid search. The coverage $c(i)$ should be in the relevant range $[g, h]$, where g and h are, respectively, the minimum and maximum coverage levels of competitors' products. We search a grid of possible coverage levels starting from the reference coverage level and going all the way to the left extreme g (the minimum coverage level offered by competitors). We check whether for the next lower coverage level in the grid the optimal portfolio still includes the insurance product, moving further down the grid until it is not included or we reach g. This gives us the minimum coverage level that the company would need to have in order to include the product in his optimal cybersecurity portfolio.

Price and coverage

Finally, we can explore price-coverage Pareto efficient insurance products by combining the two preceding methods to simultaneously optimise price and coverage, determining the maximum price that a company is willing to pay and the minimum level of coverage needed in order to include an insurance product in his optimal cybersecurity portfolio. In this case, the problem can be rewritten as

$$\arg \max_x \int u_D(x, p(i), \theta) p_D(\theta | x, c(i)) d\theta,$$

specifying the dependence of the expected utility (and the optimal cybersecurity portfolio) on both the price and the coverage, with the corresponding feasible ranges being $[a, b]$ and $[g, h]$. One approach would be to start from the reference portfolio and then present several Pareto efficient insurance products for the organisation to choose from.

Market segmentation

We now show how the model can be adapted to enable market segmentation.[2] Market segmentation involves determining clusters of organisations, defined by their organisation profile and features ft and risk aversion coefficient ρ, that would choose similar cybersecurity portfolios. This brings many benefits for insurers, including facilitating marketing operations. It also streamlines processes for them, as determining a company's optimal cybersecurity portfolio, including cyber insurance products, as shown in Section 4.3.4 is computationally intensive. Market segmentation partially alleviates this since it means that the calculations can be done once for a large group of similar organisations.

For a set of organisations characterised by their organisation profile and features and risk aversion coefficients $\{ft_i, \rho_i\}_{i=1}^m$, we can compute the corresponding optimal portfolios x_i^*, using the methods described in Section 4.3.4. Based on this, we can then find the parameters $\hat{w}$ of a metamodel $x_i^* \approx \xi(ft, \rho, w) + \epsilon$ and use $\xi(ft, \rho, \hat{w})$ to determine the portfolio that the companies would choose. We could implement the approach for the whole portfolio or for parts of it, in particular, for its cyber insurance component.

Dynamic pricing

Finally, we describe how the model can be adapted to allow for the dynamic pricing of cyber insurance.[3] Insurers can make use of a number of businesses that have emerged in recent years to provide cyber risk indicators about an organisation by aggregating information obtained from security information and event management (SIEM) or threat intelligence (TIS) systems. These systems scan the organisation's IT infrastructure, its security environment,

---

[2]Note that market segmentation is more feasible for SMEs—the main focus of this book—than for larger companies, as the larger an organisation is the more complex these calculations become.

[3]As with market segmentation, this is more practicable for SMEs.

its security posture and, whenever possible, those of its third party suppliers. (Examining the security of an organisation's third party suppliers is a new field called supply chain risk management or vendor risk management, as described in Torres et al. (2020).)

This makes it possible to define a risk index r_n for an organisation over time n and establish a benchmark risk level w so that if $r_n \geq w$ a warning is issued. Sometimes a forecasting model for r_n is introduced to facilitate predictive monitoring.

Insurers can use this to develop new cyber insurance products that are priced dynamically. For example, on top of a traditional product they could introduce a discount factor if, after a certain period, the risk indicator does not attain level w. They could also introduce a penalty if the risk indicator reaches level w and the insured company does not implement certain recommendations to improve its cybersecurity.

4.5 Cyber insurance policy issuance and fraud detection

We now present models surrounding cyber insurance fraud. As a first step, we develop a model to help an insurer decide whether or not to issue a cyber insurance policy, given the possibility that a customer might commit fraud. We describe this model in Section 4.5.3. Then, we develop a model to assist the insurer in determining whether or not to classify a cyber insurance claim by a customer as fraudulent. We describe this model in Section 4.5.5.[4]

There is a high level of insurance fraud in other areas, such as health insurance (Ekin, 2020), but the relatively low implementation of cyber insurance policies has moderated this risk in the cybersecurity field so far. However, the growth of cyber insurance and rise in the number of cyber attacks increases the likelihood of fraud, e.g. companies may be tempted to take advantage of major global cyber attack campaigns to file fraudulent claims. A new paradigm is therefore needed for fraud detection, which must efficiently detect fraudulent actions from multiple sources using large and diverse amounts of information within a relatively rapid time frame. As an example, we consider the use case of an insured company that exploits a widespread ransomware attack campaign to commit insurance fraud.

4.5.1 Use case

In order to insure against the risk of a cyber attack, a professional services company decides to take out a cyber insurance policy. The company provides advice on topics including legal and regulatory issues and business strategy. The potential impact of a cyber attack on the company includes data loss, brand damage, loss of clients, and regulatory fines. The assets that need to be protected involve customer data and business intelligence. The company does not have a high level of cybersecurity readiness. Their safeguards consist of a commercial antivirus product, a firewall for the company's internet gateway, and a data backup solution deployed internally.

Later, a widespread ransomware attack campaign occurs that infects a number of companies around the world. The professional services company is not hit by the attack, but the company CEO decides to take advantage of this in order to commit fraud. He instructs a senior IT employee to secretly make a full backup of the company data, and then to intentionally infect the company's servers with the ransomware. The company files an insurance

---

[4]Note that none of the information in either of these sections is intended to imply that customer claims are fraudulent by nature.

claim for the loss of critical business data, although it actually has a copy of this data in a secret location.

4.5.2 Current approaches

There are a number of big data analytics techniques and machine learning algorithms for detecting fraudulent claim filings, and research continues to progress rapidly in these areas. Big data analytics techniques link and process large amounts of data from a variety of sources (e.g. customer activities and behaviour, social networks, public databases, and claims history) to detect unusual patterns associated with fraud (Verma and Marchette, 2019). Machine learning algorithms employ training and validation data sets to build predictive models capable of adapting to emerging fraud behaviour, making it useful for detection as well as prevention. Robust fraud detection mechanisms integrate a combination of both big data analytics techniques and machine learning algorithms to evaluate and flag business operations susceptible to fraud using decision rules (e.g. statistical outliers, dissimilarity/suspicion scores, and risk scores). They are used to assist and complement manual checks in internal and external audits for fraud control.

These tools can be used to detect fraudulent claims involving cyber insurance as well, although the absence of sufficient cyber claims data makes their use more limited at present. In the meantime, some of the strategies that insurers can use to detect cyber claims fraud include investing in alternative quality controls and audits and accounting for fraud-related losses when pricing cyber risks.

4.5.3 Cyber insurance policy issuance: Model formulation

We now present the model that we have developed to assist an insurer in deciding whether to issue a cyber insurance policy to a potential customer, taking into account the possibility that the customer might commit fraud. We describe the insurer's decision-making process as a Bi-Agent Influence Diagram shown in Figure 4.4.

There are two agents: the *Insurer* (designated I) and the prospective *Customer* (designated J). Nodes involving just the Insurer are white and nodes involving only the Customer are gray; striped nodes are relevant to both agents. The Insurer's decision as to whether or not to *grant an insurance policy* (designated i) is modelled as a decision node. The Customer has an *organisation profile and features* (designated ft) which is modelled as a deterministic node. The *threats* (designated t) that the Customer faces are an aggregate of the different threats described in Section 4.3. These threats (and their impacts) determine the likelihood and the size of a claim, as discussed in earlier sections, and are thus modelled as an uncertainty node. They are therefore a key component of a *claim* (designated cl), which is also modelled as an uncertainty node.

Whether or not the Customer decides to commit fraud also has a major impact on claims; the *Customer's fraud decision* (designated fr) is modelled as a decision node. Should the Customer put in a claim, the Insurer (or their cybersecurity auditor) typically performs a forensic investigation regarding the claim and issues an *audit report* (designated d), modelled as an uncertainty node. If the investigation does not find any indication of fraud, then the Insurer reimburses the Customer's claim; *reimbursement* (designated r) is modelled as an uncertainty node. Both the Insurer and the Customer aim to maximise their preferences, or expected utilities. *Insurer utility* (designated u_I) and *Customer utility* (designated u_J) are both modelled as value nodes.

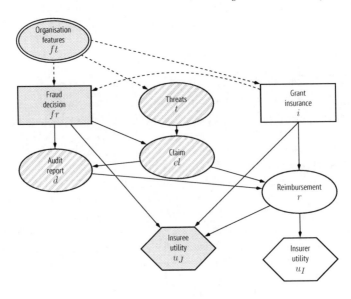

Figure 4.4: BAID for cyber insurance policy issuance

4.5.4   Model solution

The Insurer's decision is a standard ARA problem with a twist involving having to forecast whether the Customer will commit fraud, which is a strategic uncertainty. To model the Insurer's decision, we first consider the Customer. As in the Attacker problem in Section 4.3.4, we model the Customer's decision as an uncertainty and use random utilities U_J and probabilities P_J to build the Customer's expected utility model and find their (random) optimal fraud decision

$$Fr^*(i, ft) = \arg\max_{fr} \iiiint U_J(fr, i)\, P_J(d|fr, cl)\, P_J(r|d, cl, i)\, P_J(cl|t, fr)\, P_J(t|ft)$$
$$\mathrm{d}r\,\mathrm{d}cl\,\mathrm{d}d\,\mathrm{d}t.$$

We use it to assess the Customer's probability of committing fraud, from the perspective of the Insurer, as
$$p_I(fr|i, ft) = P\big(Fr^*(i, ft) = fr\big),$$
which is estimated using a Monte Carlo simulation.

This, in turn, feeds into the Insurer's preference, or expected utility, regarding whether to issue the insurance policy to the Customer

$$\psi_I(i|ft) = \iiiint u_I(r, i)\, p_I(r|cl, d)\ p_I(d|fr, cl)\, p_I(fr|i, ft)\, p_I(t|ft)\,\mathrm{d}r\,\mathrm{d}d\,\mathrm{d}fr\,\mathrm{d}t.$$

Finally, we maximise the Insurer's expected utility to decide what, if any, insurance product from a catalogue L should be offered to the Customer

$$\max_{i \in L} \psi(i|ft).$$

As with the other models presented here, this model serves as a template that can be extended or modified. For example, the model could be expanded to incorporate Customers' organisational behaviours affecting cybersecurity effectiveness (e.g. adherence to security policies or implementation of security controls). This would be useful for the Insurer when deciding whether or not to grant an insurance policy. In addition, more customer nodes could be added and the claims node could be bifurcated according to different types of claims. This would allow the Insurer to take more complexity into account. Furthermore, an adversarial threat node could be used in place of or in addition to the threat node in order to be able to assess the potential impact of a specific cyber attack on claims. This could be particularly relevant during a major cyber attack campaign like WannaCry or NotPetya.

4.5.5   Cyber insurance fraud detection: Model formulation

We now introduce the model that we have developed to assist an insurance company in determining whether it should classify a particular claim by a customer as fraudulent.[5] We model the insurer's decision-making as a Bi-Agent Influence Diagram in Figure 4.5.

There are two agents: the *Insurer* (designated I) and the *Customer* (designated F). Nodes involving just the Insurer are white and nodes involving only the Customer are gray; striped nodes are relevant to both agents. The *type of claim* (designated y) refers to whether a claim is fraudulent (designated $+$) or legitimate (designated $-$) and is modelled as an uncertainty node. In turn, the type of claim y affects the *claim features* (designated x), which also includes other characteristics such as the organisation profile and features; it is modelled as an uncertainty node as well.

Given that if the Customer commits fraud he does so by modifying the claim features x, the Customer's decision as to whether or not to commit fraud is referred to as *Customer modification* (designated m) and it is modelled as a decision node. The *modified claim features* (designated x') is the modified claim information that the Insurer receives, which is also an uncertainty node. We consider only integrity violations, i.e. cases in which the Customer's modifications are for the purposes of committing fraud. That is, we do not take accidental or erroneous modifications by the Customer into account. Based on this information, the Insurer must decide whether or not to classify the claim filed by the Customer as fraudulent; the *Insurer decision* (designated y_C) is modelled as a decision node. Both the Customer and the Insurer seek to maximise their expected utilities. *Customer utility* (designated u_F) and *Insurer utility* (designated u_I) are both modelled as value nodes.

4.5.6   Model solution

This may be seen as an ARA adversarial classification problem (Naveiro et al., 2019).

The Insurer's problem

The key elements needed to model the Insurer's decision regarding whether or not to classify a claim as fraudulent are:

- $p_I(y)$, which describes the Insurer's beliefs about the distribution of the type of claim y (that is, her beliefs about the relative proportion of fraudulent and legitimate claims);

- $p_I(x|y)$, modelling the Insurer's beliefs about the distribution of the claim features x given

---

[5]The company will typically use a classifier algorithm which examines the Customer's claim for particular characteristics and flags it if it deems there may be a chance of fraud.

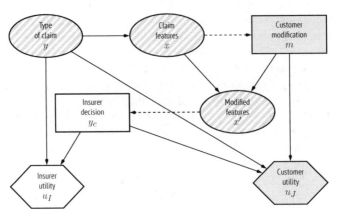

Figure 4.5: BAID for cyber insurance fraud detection

the type of claim y, when the Customer is not taken into account, thus needing $p_I(x|+)$ and $p_I(x|-)$ (that is, her beliefs about the kind of features that fraudulent and legitimate claims typically display);

- $p_I(x'|m,x)$, which models the Insurer's beliefs about the modified claim features x' given the Customer modification m and the claim features x (i.e. her beliefs about whether the claim has been modified and turned into a fraudulent claim). If we consider only deterministic transformations, it will actually be the case that $p_I(x'|m,x) = \chi(x' = m(x))$, where χ is the indicator function;

- $u_I(y_C,y)$, describing the Insurer's utility when she classifies a claim as y_C and the type of claim is y (that is, her utility when she classifies a claim correctly or incorrectly, including classifying it as fraudulent when it is legitimate and classifying it as legitimate when it is fraudulent); and

- $p_I(m|x,y)$, portraying the Insurer's beliefs about the Customer's modification m, given the claim features x and the type of claim y (that is, whether the Insurer believes the Customer has filed a fraudulent claim or not).

In addition, we assume that the Insurer can approximate the set $\mathcal{A}(x)$ of possible fraud attempts for a given claim features x. When the Insurer observes x', she can compute the set $\mathcal{X}' = \{x : m(x) = x' \text{ for some } m \in \mathcal{A}(x)\}$ of instances potentially leading to x'. She should then aim to classify the claim as either fraudulent or legitimate y_C in such a way that maximises her expected utility u_I, taking into consideration that the possibility of the Customer committing fraud modifies the probabilities $p_I(x'|y)$.

Therefore, she must consider the Customer's potential modification m of the claim features x according to the probabilities $p_I(x',x,m|y)$. In our context, this means that she must find the classification of $c(x')$ such that

$$c(x') = \arg\max_{y_C} \sum_{y \in \{+,-\}} u_I(y_C,y)p_I(y|x') =$$

$$= \arg\max_{y_C} \sum_{y \in \{+,-\}} u_I(y_C,y)p_I(y) \sum_{x \in \mathcal{X}'} \sum_{m \in \mathcal{A}(x)} p_I(x',x,m|y).$$

Furthermore, expanding the last expression and computing, we have

$$c(x') = \arg\max_{y_C} \sum_{y \in \{+,-\}} \left[u_I(y_C, y) p_I(y) \sum_{x \in \mathcal{X}'} \sum_{m \in \mathcal{A}(x)} p_I(x'|x, m) p_I(m|x, y) p_I(x|y) \right].$$

Given that we only consider modifications that are fraudulent, we have $p_I(m|x, -) = \chi(m = id)$, where id stands for the identity attack that leaves the claim features x unchanged. Then, simple computations lead to

$$\begin{aligned} c(x') &= \arg\max_{y_C} \left[u_I(y_C, +) p_I(+) \sum_{x \in \mathcal{X}'} p_I(m_{x \to x'}|x, +) p_I(x|+) \right. \\ &\quad + \left. u_I(y_C, -) p_I(x'|-) p_I(-) \right], \end{aligned} \tag{4.4}$$

where $p_I(m_{x \to x'}|x, +)$ designates the probability that the Customer will try to commit fraud by transforming x into x', when $(x, y = +)$.

The above assessments are standard (Clemen and Reilly, 2013), except for $p_I(m_{x \to x'}|x, y)$, which requires the Insurer to take into account elements of the Customer's strategic thinking.

The Customer's strategic thinking

We now consider the Customer's decision-making process. We assume that the Customer aims to modify the claim features x to maximise his expected utility u_J, which is attained by making the Insurer classify fraudulent claims as legitimate.

The Customer needs to consider the Insurer's decision about whether or not to classify the claim as fraudulent y_c as an uncertainty. Suppose, for now, that we have the following information available about the Customer:

- $p_J(x'|m, x)$, describing the Customer's beliefs about how skillfully he has modified m the claim features x into the modified claims features x' that the Insurer evaluates (that is, his beliefs about whether he has convincingly made the fraudulent claim seem legitimate). As for the Insurer, we make $p_J(x'|a, x) = \chi(x' = m(x))$;

- $u_J(y_C, y, m)$, which describes the utility of the Customer when the Insurer classifies the claim as y_C, the type of claim is y, and the modification is m (that is, when the Insurer classifies a claim correctly or incorrectly, including classifying it as fraudulent when it is legitimate or classifying it as legitimate when it is fraudulent). There are some implementation costs which are reflected here as well; and

- $p_J(c(x')|x')$, which models the Customer's beliefs about how the Insurer will classify the claim when the the Insurer observes x' (that is, his beliefs about whether the Insurer will classify the claim as legitimate when she examines the claim he has filed).

We designate by $p = p_J(c(m(x)) = +|m(x))$ the probability of the Customer admitting to the Insurer that he has filed a fraudulent claim, given that she discovers the falsification $x' = m(x)$. Since she will have uncertainty about the probability of this occurring, we denote its density by $f_J(p|m(x))$ with expectation $p^J_{m(x)}$.

Among the different possible modifications m, the Customer would choose that maximising his expected utility

$$m^*(x, y) =$$
$$= \arg\max_m \int \left[u_A(c(m(x)) = +, y, m) \cdot p + u_J(c(a(x)) = -, y, m) \cdot (1 - p) \right] f_J(p|m(x)) dp. \tag{4.5}$$

As we assume that the Customer does not modify the claim when it is legitimate, we only consider the case in which $y = +$. Then, the Customer's expected utility when he engages in modification m and the situation is $(x, y = +)$ will be

$$\int \left[u_J(+, +, m)\, p + u_J(-, +, m)\, (1 - p) \right] f_J(p|m(x)) dp =$$
$$= \left[u_J(+, +, m) - u_J(-, +, m) \right] p^J_{m(x)} + u_J(-, +, m). \quad (4.6)$$

Maximising the Insurer's expected utility given the Customer's strategic thinking

However, the Insurer does not know the utilities u_J and expectations $p^J_{m(x)}$ of the Customer. We model her uncertainty through a random utility function U_J and a random expectation $P^J_{m(x)}$. We can then solve for the random optimal modification, optimising the random expected utility

$$M^*(x, +) = \arg\max_m \left(\left[U_J(+, +, m) - U_J(-, +, m) \right] P^J_{m(x)} + U_J(-, +, m) \right),$$

and make $p_I(m_{x \to x'}|x, +) = Pr(M^*(x, +) = m_{x \to x'})$, assuming that the set of modifications is discrete. We use a Monte Carlo simulation to estimate such probabilities. This would feed problem (4.4) which can now be solved.

An operationalisation of the above framework in the context of spam detection may be seen in Naveiro et al. (2019).

4.6 Cyber reinsurance decisions

Finally, we develop a model to help insurers determine the level of cyber reinsurance they should take out. The model also makes it possible to better understand accumulation risk, which is the risk that a single event spreads to multiple lines of business, leaving insurers with unexpectedly large losses.

The use case addresses risk, looking at a large-scale cyber attack that has major repercussions on numerous sectors, resulting in an unusually high number of insurance claims. To protect against the risk of an extreme cyber event, insurers typically purchase reinsurance for excess losses (i.e. losses beyond a threshold amount), partially transferring the risk to one or several reinsurance companies.

4.6.1 Use case

An organised crime group launches a major ransomware attack campaign that compromises large numbers of computers and other devices. The non-targeted nature of the campaign combined with a high number of vulnerabilities and unpatched devices means that the attack infects numerous companies in a wide range of market sectors and segments, causing significant data loss for these companies (due to being encrypted by the ransomware).

Moreover, the attack also impacts companies providing telecommunications and cloud services, thereby interrupting communication channels and affecting the business continuity of other companies that depend on these services as well. This adds substantial severity to the impact of the attack.

For all of these companies, the loss of critical data (despite remediation mechanisms such as backed up data, redundancy, and crisis management plans) and/or impact on business continuity entail significant financial losses. These companies all address claims to their respective insurance companies. Because many market sectors and segments are impacted, the attack results in claims for large portions of insurance companies' portfolios. However, this can be lowered to acceptable levels due to reinsurance decisions.

4.6.2 Current approaches

Cyber reinsurance is critical in order to deal with accumulation risk, as it allows insurers to partially transfer the risk of catastrophic cyber losses to reinsurers. Reinsurers struggle with the same challenges that insurers do when underwriting insurance policies involving cybersecurity: namely, pricing problems stemming from the lack of cyber incident data and the constantly evolving nature of cyber risk. These uncertainties have constrained the development of cyber reinsurance thus far. Reinsurers have tended to take a conservative approach, offering mainly standalone affirmative cyber risk cover on a proportional basis, with annual aggregate limitations. As the industry acquires greater cyber incident data over time, however, the provision and affordability of cyber reinsurance is expected to grow.

In addition, cyber insurers can obtain financing for extreme cyber risks in the capital markets. Instruments such as catastrophe bonds (cat bonds) can be an alternative to reinsurance for major cyber insurers: risk financing with cat bonds could be useful in order to cover potentially catastrophic risk (e.g. a major electric grid breach that could cause trillions in cascading losses).

Given that reinsurance for critical infrastructure is very limited, governments should also consider acting as reinsurers of last resort (Woods and Simpson, 2017). Finally, by using financial incentives to encourage companies to improve their cybersecurity readiness, reinsurance and cyber insurance will make insurers more resilient to catastrophic cyber risks.

4.6.3 Model formulation

We now describe the model developed to assist an insurer in deciding on the level of reinsurance to take out against cyber attacks. It also helps an insurer better understand accumulation risk as follows: In the model, the insurer has engaged in market segmentation (as discussed in Section 4.4.3), having developed different types of insurance products for different market segments. By detailing the individual market segments in the model, it is possible to isolate and analyse the accumulation effect from a cyber attack affecting a given market segment. We represent an insurer's cybersecurity reinsurance decision as an Influence Diagram in Figure 4.6.

In this model we have just one agent, the *Insurer*, who is the one possibly taking out the reinsurance. There are three market segments:

- Standard SMEs, which we refer to as *Segment 1* (designated S_1). These companies have very limited IT staff; in some cases they may not have any.

- IT-intensive SMEs, which we refer to as *Segment 2* (designated S_2). These companies consider IT to be a critical or core function for their business. They employ dedicated IT teams.

- Large enterprises, which we refer to as *Segment L* (designated S_L). These companies maintain major IT infrastructures and usually have large in-house IT departments and other security functions.

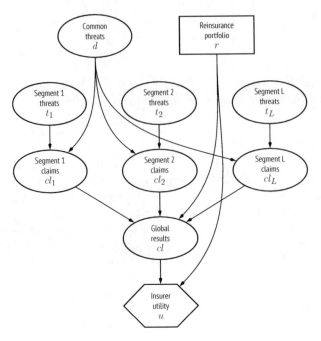

Figure 4.6: Influence Diagram for cybersecurity reinsurance

These market segments face various types of threats. *Common threats* (designated d) affect all market segments; that is, they are common to all companies. There are also market segment-specific threats: *Segment 1 threats* (designated t_1) primarily affect standard SMEs; *Segment 2 threats* (designated t_2) primarily affect IT-intensive SMEs; and *Segment L threats* (designated t_L) primarily affect large enterprises. For example, malware attacks that are untargeted may affect all companies; however, social engineering attacks carried out by competitors may primarily target IT-intensive SMEs or large enterprises. These threats are all modelled as uncertainty nodes.

These threats have an impact on claims. *Segment 1 claims* (designated cl_1) are claims made by standard SMEs; *Segment 2 claims* (designated cl_2) are claims made by IT-intensive SMEs; and *Segment L claims* (designated cl_L) are claims made by large enterprises. These are also modelled as uncertainty nodes.

Claims, in turn, affect *global results* (designated cl), which are the Insurer's profits. It is modelled as an uncertainty node. The *reinsurance portfolio* (designated r) describes the reinsurance portfolio options that the Insurer has to choose from and is modelled as a decision node. These options could involve products from several reinsurers. They could also be affected by financial, legal, or compliance requirements. The Insurer seeks to maximise his utility; *Insurer utility* (designated u) is modelled as a utility node.

4.6.4 Model solution

Having engaged in market segmentation, for each market segment we consider its size n_i and the aspects of a typical insured company such as their IT systems, cybersecurity readiness, financial resources, features, assets, and threats, as described in Section 4.3.4. (Insured companies in the same segment can be assumed to be positively correlated with correlation ρ_i, so that a typical insured company in each segment can be established.) Based on this, each insured company j in the i-th segment is assumed to produce claims with distribution $f_j(cl_j)$, pay a premium p_j, and receive coverage r_j in the event of a claim.

The cl_1, cl_2, and cl_L nodes summarise all of this for each segment. For instance, in the case of companies belonging to Segment 1, or standard SMEs, the cl_1 distribution depends on n_1, ρ_i, $f_1(cl_1)$, and r_1. The cl node aggregates the effects of cl_1, cl_2, and cl_L over various segments, yet is compensated by the reinsurance decision r so that $cl = g(cl_1, cl_2, cl_L, r)$. The reinsurance decision could be restricted by various financial, legal, or compliance requirements and it might also involve a portfolio of several reinsurers.

Once we are capable of building $p(d)$, $p(t_i)$, $p(cl_i|t_i, d)$, $i = 1, 2, L$ and the utility function u_I for the insurance company, we maximise

$$\max_r \int \cdots \int u_I\big(g(cl_1, cl_2, cl_L, r), r\big)\, p(d) \prod_i p(t_i) \prod_i p(cl_i|t_i, d)\, \mathrm{d}cl_1\, \mathrm{d}cl_2\, \mathrm{d}cl_L\, \mathrm{d}t_1\, \mathrm{d}t_2\, \mathrm{d}t_L,$$

to find the optimal reinsurance decision for the insurance company.

In addition, we can use this model to study the impacts of accumulation risk. By having separated out the threats according to the market segments that they affect, we can assess the accumulation effect of an attack on a particular market segment.

As with the models presented in previous sections, this model is a basic template and can be extended further. The number of market segments could be increased or adapted. For example, they could be differentiated by sector (e.g. electric power companies) or by country. In addition, the number of common and specific threats could be expanded or modified as well.

4.7 Conclusions

This chapter has set forth potential innovative solutions to some of the risk management challenges for organisations and insurance companies when it comes to cybersecurity and cyber insurance that were described in the first three chapters of this book. We presented the CSRM model to assist organisations in their cybersecurity resource allocation decisions, including cyber insurance product selection (Section 4.3). We also presented a series of auxiliary models to assist insurance companies. These included models for cyber insurance product design (including dynamic pricing) (Section 4.4), cyber insurance policy issuance and fraud detection (Section 4.5), and cyber reinsurance decisions (including better understanding accumulation risk) (Section 4.6). We made use of ARA and MAIDs in developing these models, improving considerably on previous approaches by taking adversarial aspects into account. These models all stem from relevant use cases in order to ensure that the models are grounded in practical experience.

The models that we presented here are basic templates that can be expanded, reduced, or otherwise modified. They can also be interconnected. For example, the models for cyber insurance product design in Section 4.4 can be used to update the information in the CSRM cybersecurity resource allocation model in Section 4.3; specifically, the variables price $p(i)$ and coverage $c(i)$ which we seek to optimise are parameters of the insurance i decision node in the CSRM model. As another example, the cyber insurance policy issuance and fraud detection models in Section 4.5 may be used to update the information in the Section 4.3 CSRM model by embedding both models.

Enabling dynamic pricing of insurance products

One of this chapter's major contributions is in looking at how to enable the dynamic pricing of insurance products. This allows insurers to model the risk to an insured company much more accurately at any given time, leading to benefits for both insurers and insured companies by making it possible to price insurance products more accurately. As discussed in Section 4.4, insurers can now take advantage of a growing number of businesses that are able to scan an insured company's IT infrastructure, the IT infrastructure of its suppliers, and obtain other indicators about the company's security posture. This makes it possible to create a risk index for the insured company that is automatically adjusted in response to a change in any of these factors, and insurers can use this information to adjust the insured company's premium dynamically.

In addition to a company's security posture, the dynamic pricing technique could be extended to take other dynamic aspects of cybersecurity into account. It could attempt to capture the rapidly evolving nature of cyber attacks. Current cyber attack types are growing in size and sophistication (e.g. DDoS attacks leveraging the power of millions of compromised IoT devices), new attack types frequently emerge (e.g. AI-enabled attacks), and new attackers regularly appear on the scene (e.g. more nation states developing advanced cyber attack capabilities). Certain sectors also experience surges in cyber attacks at various times, when they are targeted by major cyber attack campaigns (e.g. a wave of ransomware attacks against hospitals or of malware attacks directed at the energy sector). These changes too could be captured with various cybersecurity risk indices and used to dynamically adjust premiums.

The dynamic nature of cybersecurity also means that it is important to regularly revisit and rerun the models presented here. For example, in the case of the fraud detection model in Section 4.5, we would expect fraudsters to modify their attack techniques so that the calculations would need to be periodically reevaluated.

Better understanding cyber accumulation risk

Another significant contribution of this chapter has been to make it possible to better understand the accumulation risk from cyber. Accumulation risk can lead to substantial losses for insurers and can threaten their bottom line. It poses an even greater risk for cyber insurance (compared to other types of insurance) due to the lack of geographic boundaries in the cyber realm: In contrast to an earthquake or tornado whose effects are confined to a specific region, a cyber attack can inflict damage around the world. These factors all make the use of the model we have developed to better understand cyber accumulation risk all the more important. As explained in Section 4.6, we can do so by detailing individual market segments in the model, making it possible to isolate and analyse the accumulation effects from a cyber attack affecting a given segment.

Leveraging the prototype toolbox to enhance cybersecurity

These models could be implemented in a decision support system, i.e. a computerised system that supports businesses in their decision-making activities or processes. As part of the CYBECO project, we developed and tested such a decision support system in the form of a prototype "toolbox" based on the CSRM model that provides an online interface for company decision makers who wish to use it, as described in Chapter 3. We have also made this publicly available for companies and others to use. In response to information inputted by the company, the toolbox outputs an assessment of its cybersecurity risk as well as advice on how they can best allocate their cybersecurity resources between purchasing cybersecurity products and cyber insurance. In Chapter 3 we used a Behavioural Economics Experiment (BEE) to test the toolbox. We examined both how its use impacted subjects' decision-making regarding cybersecurity and cyber insurance and also studied how changes to the display page could nudge better cybersecurity and cyber insurance decisions. The use of such a decision support system by companies can significantly simplify the decision-making process for them. It can also encourage the uptake of better cybersecurity measures and the adoption of cyber insurance, improving the cybersecurity of the ecosystem as a whole. This is described in more detail in Couce-Vieira et al. (2020b).

Appendix

This section provides a simple numerical example to illustrate the concepts introduced in this chapter. We consider a simplified version of the CSRM problem presented in Section 4.3 and present two different approaches: The first one involves carrying out a standard risk analysis. The second one makes use of Adversarial Risk Analysis, taking into account that one of the threats is due to an adversarial Attacker.

Use case

An SME that provides services over the internet must decide on its cybersecurity strategy for the coming year. There are two potential threats that could impact the company: a virus and a DDoS attack. To defend against these threats, the SME could purchase one or both of the following security controls: a malware protection system (which makes infection by a virus less likely) and/or a DDoS mitigation system (which makes a DDoS attack less likely and less effective). In addition, the company could also buy a cyber insurance product.

The SME must therefore decide which security controls to purchase and whether or not to purchase the cyber insurance product. Its annual cybersecurity budget is $7,000. The company is risk neutral and thus aims to minimise its expected cybersecurity costs.

Standard risk analysis approach

We depict a standard risk management approach to the CSRM problem using an Influence Diagram in Figure 4.7.

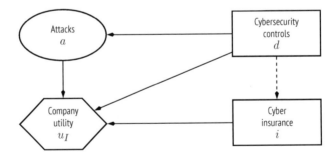

Figure 4.7: Standard risk management approach to the CSRM problem (without ARA)

Table 4.1 summarises the probabilities of various threats occurring depending on whether or not the security controls are implemented, as well as the costs of the security controls and of the cyber insurance product. For example, if we implement both security controls, there is only a 0.01 probability of a successful DDoS attack in the coming year and the cyber insurance premium is $800. The DDoS mitigation system costs $4,000 and the malware protection system costs $2,000. The threats are assumed to be independent.

	MPS	DDMit	Both	None
Virus	.05	.10	.05	.10
DDoS	.05	.01	.01	.05
Both	.0025	.001	.0005	.005
Cost	2	4	6	0
Cyber insurance cost	1	1	0.8	3

Table 4.1: Probabilities of various threats occurring depending on the security controls implemented, as well as costs of the security controls and of cyber insurance, in thousands of dollars

Table 4.2 shows the expected impacts of various attacks, depending on the security controls that have been implemented. For example, if we implement the DDoS mitigation system and the company is infected with a virus, then the expected cost to the company is $35,000.

We can observe the effects of various security controls on the probabilities (Table 4.1) and/or impacts (Table 4.2) of the attacks.

	MPS	DDMit	Both	None
Virus	35	35	35	35
DDoS	100	10	10	100
Both	135	45	45	135

Table 4.2: Expected costs of attacks, in thousands of dollars

The cyber insurance product covers 50% of the costs due to an attack, with a maximum annual cap of $60,000.

We now compute the expected costs of the cybersecurity portfolios. As an example, the cybersecurity portfolio (MPS and cyber insurance) has an expected cost of

$$\psi(MPS, cyber insurance) = .05 \times (0.5 \times 35) + .05 \times (0.5 \times 100)$$
$$+ .0025 \times (135 - \max(0.5 \times 135, 60)) + .8975 \times 0 + 3 = 6.5625.$$

Table 4.3 displays the expected costs of the cybersecurity portfolio, as well as the costs of the security controls.

Security control	Insurance	Expected cost	Cost to purchase
None	No	9.175	0
MPS	No	10.125	2
DDMit	No	7.735	4
Both	No	8.075	6
None	Yes	7.625	3
MPS	**Yes**	**6.5625**	**3**
DDMit	Yes	6.8225	5
Both	Yes	7.78625	6.8

Table 4.3: Effectiveness and costs of the cybersecurity portfolios, in thousands of dollars

All of the cybersecurity portfolios are feasible as their costs all fall within the cybersecurity budget. In this case, we would select the cybersecurity portfolio containing the MPS and the cyber insurance product, as it has the lowest expected cost.

Adversarial Risk Analysis approach

We now consider a potential targeted attack from an adversary, as the SME has learned that one of its competitors might be considering hiring a third party to attack it using a DDoS attack

This problem is depicted as a BAID in Figure 4.8. The white nodes refer to the SME (the Defender) and the gray nodes refer to the competitor (the Attacker). The striped nodes refer to events that are relevant to both agents. (In this case, the θ node refers to whether or not the DDoS attack is successful. This depends on the security controls that have been implemented.) When deciding whether or not to wage a DDoS attack against the Defender, the Defender has knowledge of the security controls that the Defender has implemented, as indicated by the arrow from the Defender decision node to the Attacker decision node.

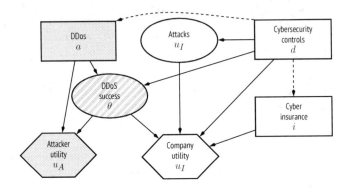

Figure 4.8: ARA approach to the CSRM problem

We analyse the Attacker's strategic thinking, for which we assume the random assessments in Table 4.4. For example, if the SME implements the DDoS mitigation system, the expected utility is modelled with a beta $\beta e(9,1)$ distribution (mean $9/10$) when the attack is successful and with a beta $\beta e(1,100)$ distribution (mean $1/101$) when it is unsuccessful; the probability of success is modelled with a $\beta e(1,10)$ distribution (mean $1/11$). We scale utilities between 0 and 1.

Security control	Utility if successful	Utility if unsuccessful	Prob
DDMit	$\beta e(9,1)$	$\beta e(1,100)$	$\beta e(1,10)$
No DDMit	$\beta e(14,1)$	$\beta e(1,90)$	$\beta e(10,10)$

Table 4.4: Random assessments for the Attacker

We simulate from the Attacker's strategic thinking in order to obtain the probabilities of various attacks. Simulating the attacking process 1,000 times, we obtain the probabilities of various threats occurring depending on whether or not the security controls are implemented, shown in Table 4.5. As we can see, if the DDoS mitigation system is implemented, the probability of a DDoS attack is approximately 0.001, whereas if it is not implemented, the probability is approximately 0.4. Thus, the DDoS mitigation system has a clear deterrent effect. (NB: The costs of the security controls and of the cyber insurance product are the same as those provided in Table 4.1.)

	MPS	DDMit	Both	None
Virus	.05	.10	.05	.10
DDoS	.4	.001	.001	.4
Both	.02	.0001	.0005	.04

Table 4.5: Probabilities of various threats occurring depending on the security controls implemented

Table 4.6 shows the expected costs of the cybersecurity portfolios. (NB: The costs to purchase the cybersecurity portfolios are the same as those provided in Table 4.3.)

Security control	Insurance	Expected cost
None	No	44.04
MPS	No	44.71
DDMit	No	7.67
Both	No	7.92
None	Yes	27.75
MPS	Yes	25.37
DDMit	**Yes**	**6.81**
Both	Yes	7.76

Table 4.6: Effectiveness of the cybersecurity portfolios, in thousands of dollars

Thus, in this case, we would select the cybersecurity portfolio containing the DDoS mitigation system and the cyber insurance product, as it has the lowest expected cost.

5

A Case Study in Cybersecurity Resource Allocation and Cyber Insurance

Aitor Couce Vieira, David Ríos Insua, Alberto Redondo

ICMAT

Caroline Baylon

AXA

CONTENTS

This chapter presents a case study to illustrate the Cybersecurity Risk Management model developed in Chapter 4, which is designed to assist organisations in selecting their optimal "cybersecurity portfolio," including their choice of a cyber insurance product. The case study demonstrates how to implement the model and is accompanied by detailed numerical examples. We cover the formulation of the model, how to assess each of the components, and how to solve for the optimal cybersecurity portfolio. This includes showing how to assess components in situations in which there is limited data, relying on structured expert judgement techniques instead. Our aim is for this case study to serve as a blueprint for companies or other researchers who wish to use the model we have developed. To facilitate this, we have made the R code used to run the model publicly available.

5.1 Introduction

In Chapter 4 we developed a series of models to assist organisations and insurers with their decisions involving risk management in cybersecurity. In particular, we developed a key model to support companies in their Cybersecurity Risk Management (CSRM) decisions in Section 4.3 which we refer to as "the CSRM model". The model helps organisations determine their optimal "cybersecurity portfolio," or the best way for them to allocate their budget between spending on security controls to defend against cybersecurity threats and on cyber insurance products to indemnify them if they are attacked. In this chapter we set forth a detailed case study illustrating how to implement the CSRM model, along with detailed numerical examples. We cover its formulation, how to model each of the components, and how to solve for the optimal cybersecurity portfolio. This builds upon the work of Rios Insua et al. (2019) and Couce-Vieira et al. (2020b).

In addition, as part of this case study we assess various components of the model in situations in which there is limited data, relying on expert judgement instead. The use of structured expert judgement techniques, as described in Cooke and Goosens (2000), typically involves holding individual expert elicitation sessions in which they can share their views, ideally based on written expert rationales they have drafted beforehand. They also generally include a preparation for elicitation phase to identify the key issues and experts, sometimes conducting a dry run exercise and an expert training session, as well as a post elicitation phase where the views of the different experts are combined and analysed.

Our hope is that this case study will serve as a blueprint for companies or other researchers who wish to use the model we have developed. They can extend it by taking into account different or additional assets, threats, impacts, security controls, and other factors, depending on the characteristics of the company they are modelling. To make it easier for them to adapt it for their particular needs, we have also made the R code used to run the model publicly available. It can be accessed at `https://github.com/cybeco/cybersecandcyberinsurance.git`.

5.2   Case description

5.2.1   Case study

This case study focuses on Median, an IT-intensive SME that provides web hosting and web application services in the country of Small Nation. Its customers include several of the country's largest companies that are critical to the national economy. It controls 50% of the market, whose estimated annual value is $15 million.

Median believes that there are four main threat actors that are likely to try to attack it. This includes Compeet, Median's major competitor with whom it vies for market share. The hacktivist organisation Antonymous, which has criticised Median for some of its business dealings, could target Median as well. Median is also concerned about Cybegangsta, a major cybercriminal organisation that has recently emerged. Finally, the country of Modern Republic may seek to attack Median as a means to harm Small Nation, with whom it has long had tensions. In addition, Median has to consider other threats to the security of its IT systems, including environmental and accidental threats as well as untargeted cyber attacks that infect companies indiscriminately.

The SME is concerned about a major cyber attack that could prevent it from delivering services or that could result in a major data breach. The company possesses sensitive data, including some 266,000 data records. Of these, 200,000 records consist of information about its clients or employees, much of it Personally Identifiable Information (PII). The remaining 66,000 records involve confidential business information, notably surrounding the company's business activities.

In terms of assets, Median has an IT installation composed of 300 computers and 40 servers. Each computer is valued at $400 and each server at $1,200. It also has an off-site backup system. The company has 50 employees. It already has an IT team in place with appropriate training and experience and no new hires are expected. Median rents its office space.

The company has a cybersecurity budget of $72,000 for the coming year. It needs to determine how to allocate this budget between spending on security controls and on cyber insurance. It already complies with the Small Nation Cyber Essentials scheme, which recommends that all companies have at least two basic security controls: (1) a firewall and internet gateway and (2) an access control system. Following the suggestion of Cyberco, a cybersecurity consultant that the company has hired, Median is considering investing in a number of additional security controls and must determine which ones to purchase. Median has a choice between several insurance products. This includes a conventional insurance product and two cyber insurance products, one providing a more basic level of cyber coverage and another offering more comprehensive cyber cover.

Given that companies that are key to the national economy depend on Median for important services, the national regulator classifies Median as a critical national infrastructure and asks it to undertake an in-depth cybersecurity risk assessment. This requires Median to go beyond the basic approaches currently used in Small Nation, such as those based on MAGERIT (Amutio et al., 2012).

5.2.2 Preliminary assessment

Median has requested our assistance in carrying out this in-depth cybersecurity risk assessment. Working with Median's management and IT teams, we conduct a preliminary assessment of the company's cybersecurity risk, which is presented in this section. The process consists of the following steps:

- identifying the company's assets;

- identifying the unintentional threats that affect the company. These include environmental threats, accidental threats, and non-targeted cyber threats;

- identifying the intentional/targeted cyber threats that affect the company. This involves determining the company's key adversaries as well as the types of attacks that these adversaries are most likely to carry out, given their capabilities and objectives;

- considering the potential impacts of these threats on the company's assets;

- identifying possible additional security controls that the company could implement. This includes considering how these security controls would affect the likelihood and impact of these threats occurring;

- identifying potential insurance products that the company could purchase;

- identifying the decision factors that adversaries take into account when deciding whether or not to attack the company. This includes their uncertainties and their objectives; and

- determining the relevant constraints.

To facilitate the tasks above, there are a number of threat, asset, and security control catalogues available, many of them put out by organisations that have developed some of the current cybersecurity risk assessment methodologies described in Section 1.1.

Assets

We start by identifying the company's key assets. These include the company's data records, which involve (i) databases containing customer and employee data, including PII, and (ii) confidential business information about the company. Other key assets include the company's computer equipment and IT systems as well as its market share. This is summarised in Table 5.1. Note that the premises are not included as assets, since they are rented.

Assets	Description
Customer and employee data	Customer and employee data, including PII, stored on company computers
Business information	Confidential business information belonging to the company
IT systems and computer equipment	Data centre, including servers, and employee computers
Market share	Percentage of national market controlled by the company

Table 5.1: Key assets for Median

To simplify this case study, we limit ourselves to the company's top four assets. However, others using this model may want to include additional assets, depending on the characteristics of the company they are modelling. This might involve the company's reputation, its installations if owned, its stock market valuation, or its intellectual property, such as proprietary software.

Unintentional threats

We next consider the major unintentional threats that could impact these assets, by which we mean threats not specifically intended to cause harm to Median. As mentioned above, these can be divided into environmental, accidental, and non-targeted cyber threats, according to the ISF's classification described in Chapter 1 (ISF, 2017). For the purposes of this case study, we limit ourselves to the top two threats impacting Median in each of these three categories.

Environmental threats

Environmental threats are threats outside the control of an organisation that can cause harm to its IT systems; they might involve natural hazards, man-made hazards, or failures of critical infrastructure (ISF, 2017). The top two environmental threats that could impact Median's IT systems are a fire and a flood. These are shown in Table 5.2.

Environmental threats	Description
Fire	Fire damage to Median's IT systems
Flood	Flood damage to Median's IT systems

Table 5.2: Environmental threats impacting Median

For others using this model, environmental threats such as earthquakes, hurricanes, or tornadoes may also be relevant, depending on where the company they are modelling is headquartered or has operations. They may also wish to consider power failures or disease outbreaks.

Accidental threats

Accidental threats are threats that as a result of error or unintentional action (or lack of action) cause harm to an organisation's IT systems (ISF, 2017). The top two accidental threats that can affect Median are employee error and misconfiguration. These are shown in Table 5.3. For the purposes of this case study, we define employee error relatively narrowly as an employee's unintentional disclosure, modification, or deletion of company data records (for example, by accidentally emailing an attachment to the wrong person, storing data records in an insecure location, or losing a laptop containing company data records). Misconfiguration consists of the IT team incorrectly configuring software or hardware. (There have been a number of data breaches due to misconfigurations of cloud databases in recent years.)

Accidental threats	Description
Employee error	Unintentional disclosure, modification, or deletion of data records by an employee
Misconfiguration	Accidental misconfiguration of software or hardware by the IT team

Table 5.3: Accidental threats impacting Median

Other organisations who wish to use this model may want to consider additional accidental threats. For example, this might include an employee falling for a phishing attack. The degree to which the company being modelled is susceptible to such a threat will depend on such factors as the level of cybersecurity awareness within the company (e.g. the extent of cybersecurity awareness training that employees have undergone).

Non-targeted cyber threats

Non-targeted cyber threats are those that do not target a specific company, aiming to infect as many as possible. The two main non-targeted cyber threats impacting Median are ransomware and viruses. These are shown in Table 5.4. Ransomware, which acts by encrypting a company's data records until a ransom payment is made, would stymie Median's ability to operate. A virus, or computer program that replicates itself by modifying other computer programs and inserting its own code, could similarly leave Median unable to deliver services.

Non-targeted cyber threats	Description
Ransomware	Encryption of the company's data records
Virus	Infection of the company's IT systems and computer equipment

Table 5.4: Non-targeted cyber threats impacting Median

Others using this model may want to consider additional non-targeted cyber threats such as waterholing, scanning, or phishing, as relevant.

Intentional/targeted threats

In contrast to non-targeted cyber threats, an intentional or targeted threat involves a cyber attack that is aimed at a specific company. Note that when it comes to distinguishing between non-targeted and targeted threats, some types of attack methods are used in both instances. For example, some ransomware hits companies indiscriminately, while other ransomware might be targeted at a specific sector such as healthcare or infrastructure.

Key adversaries

We first identify Median's key adversaries. For simplicity, we limit ourselves to the top four potential attackers in this case study, which were briefly introduced in Section 5.1. These include Compeet, a major competitor with whom Median vies for control of the national market. Compeet may try to attack Median in order to disrupt Median's ability to deliver services. This would cause Median reputational harm, enabling Compeet to gain market share from Median. The hacktivist organisation Antonymous, which has publicly criticised Median due to some of the companies that it provides web hosting services to, could try to launch a cyber attack campaign against Median in retaliation. The cybercriminal organisation Cybegangsta is interested in attacking Median for financial gain. This could involve stealing its data, including PII, to sell to other cybercriminals. Finally, Modern Republic, which has long had tensions with Small Nation, may seek to attack Median as part of its activities directed against it. It could be to disrupt the business operations of some of the country's major companies that rely on Median's services, causing economic harm to Small Nation. Or it might be to steal information that Median holds about some of these companies, either as part of its intelligence-gathering operations or to gain economic advantage. These adversaries are detailed in Table 5.5.

Attackers	Description
Compeet	Motivation is to cause Median reputational harm, enabling it to gain additional market share
Antonymous	Seeks retribution against Median for some of the company's business dealings
Cybegangsta	Goal is financial gain
Modern Republic	Views attacking Median as a means to sabotage Small Nation, by causing harm to major companies that are customers of Median

Table 5.5: Adversaries interested in attacking Median

Others using this model may wish to consider additional potential attackers. These might include terrorist groups; those hacking for "recreational" purposes, motivated by a desire for fame and notoriety; or disgruntled former employees, who might wish to elicit revenge on the company. This will depend on the particular circumstances of the company being modelled.

Types of targeted threats

In terms of the types of targeted attacks that these attackers might perpetrate, they include DDoS attacks and social engineering attacks. DDoS attacks involve flooding a company or other target with internet traffic, overwhelming it so that it is unable to deliver services. Social engineering attacks involve the use of deception to manipulate individuals into divulging confidential information that can be used for malicious purposes. They might include the use of spear phishing (phishing attacks targeted at a particular individual), manipulating individuals inside a company into disclosing sensitive information or granting them unauthorised access, or even planting subversive individuals inside the company. These are listed in Table 5.6.

For the purposes of this case study, we choose the top two targeted threats impacting the company. Others using this model may want to consider additional targeted threats, such as supply chain attacks or the insider threat. They may also wish to consider physical

Targeted threats	Description
DDoS attack	Flooding Median with traffic, taking down Median's services and also impacting the companies that depend on it
Social engineering	Manipulation of Median employees into disclosing passwords or other confidential information and/or other attempts to gain unauthorised access

Table 5.6: Targeted threats impacting median

ones such as bombs. Again, the most relevant targeted threats will depend on the profile of the company being modelled.

Types of targeted attacks pursued by each attacker

Some attackers will be more likely to engage in certain types of attacks than others. Based on our knowledge of these attackers' interests and capabilities, we assess what types of attacks they tend to perpetrate. This is reflected in Table 5.7. For example, Compeet appears to be primarily interested in and capable of carrying out DDoS attacks, most likely by hiring a third party.

Attack	Compeet	Antonymous	Cybegangsta	Modern Republic
DDoS	x	x		x
Social engineering			x	x

Table 5.7: Types of attacks carried out by each attacker

Threat impacts on assets

We next consider the potential impacts of these threats on the company's assets. These are identified as:

- damage to *IT systems and computer equipment* (denoted as *Equip*);

- loss of *market share* to Median's main competitor (denoted as *MShare*);

- loss of *availability*, i.e. inability to deliver services (denoted as *Avail*). This has a direct impact on market share. The assessed cost per hour of downtime is $60,000;

- exposure and/or loss of *customer and employee data, including PII* (denoted as *CustEmpData*);

- exposure and/or loss of *business information* (denoted as *BusInfo*); and,

- *fines* that must be paid to the regulators in case of a major data breach (denoted as *Fines*). This is closely related to the impact on customer and employee data, including PII.

For this case study, we have selected the top six threat impacts on the company. However, others wishing to use this model may want to consider additional impacts. These include reputational damage and the related concept of brand damage, as well as a fall in the company's share price, among others.

Impacts of each individual threat on assets

Examining these threats in further detail, for each individual threat we look at the impact(s) that it can have on assets. This is shown in Table 5.8. For example, a fire may damage IT systems and computer equipment; however, it is unlikely to cause a significant loss of availability due to the company having an external backup system.

Threats	Equip	MShare	Avail	CustEmpData	BusInfo	Fines
Fire	x					
Flood	x					
Employee error				x	x	x
Misconfiguration	x			x	x	
Ransomware			x	x	x	
Virus	x		x	x	x	
DDoS		x	x			
Social engineering				x	x	x

Table 5.8: Threat impacts on assets for each individual threat

Security controls

Median already has certain security controls in place and it must decide which additional security controls to purchase for the coming year. Those implemented correspond to the security controls needed to comply with the Small Nation Cyber Essentials scheme. These are:

- a *firewall and internet gateway* (denoted as *FwGw*), and

- an *access control system* (denoted as *ACS*).

The additional security controls that Median is considering have been proposed by the cybersecurity consultant Cyberco and include a mix of physical security controls and IT security controls. These involve:

- a *sprinkler system* (denoted as *Sprk*);

- the implementation of a *flood door* and flood risk management procedures (denoted as *FD*);

- a *DDoS mitigation system* (denoted as *DDMit*);

- the use of a *secure configuration* (denoted as *SecCnf*);

- a *malware protection system* (denoted as *MPS*);

- a *patch vulnerability management system* (denoted as *PVM*); and

- an *intrusion detection system* (denoted as *IDS*).

Others using this model may wish to consider additional security controls based on the characteristics of the company they are modelling. This might include encryption of data records or the use of honeypots. They may also want to include physical security controls such as barbed wire.

Effect of security controls on the likelihood and impact of threats

We next examine how the proposed security controls affect the likelihood of the various threats occurring and/or their impact if they do occur. This is displayed in Table 5.9. For each security control, if we determine that it reduces the likelihood (denoted as L) of a given threat occurring and/or its impact (denoted as I), then an L and/or I is placed in the corresponding box. For example, the I in the (Fire, Sprk) entry indicates that if the sprinkler system is implemented, a fire will have less of an impact; however, there is no L because it does not reduce its likelihood.

Of course, the price of these security controls is an important consideration for Median, and the table also indicates the annual cost of each. For the security controls that Median has already put in place (FwGw and ACS), these represent the annual operational costs. For the security controls that Median is considering implementing, these are a combination of the initial purchasing costs and the annual operational costs.

Threats	FwGw	ACS	Sprk	FD	DDMit	SecCnf	MPS	PVM	IDS
Fire			I						
Flood				I					
EmpErro		L							
Misconfig		L,I				L,I			
Ransomware	L					L	L	L	L
Virus	L					L	L	L	L
DDoS					L,I				
SocialEng	L	L							L
Cost	$5,600	$6,000	$600	$4,800	$12,000	$1,000	$4,000	$1,600	$30,000

Table 5.9: Effect of security controls on the likelihood (L) and/or impact (I) of threats, and their costs

Insurance products

Median has a choice between the following insurance products for risk transfer purposes:

- a *conventional insurance product* for property damage (denoted as *Conv*) that partly covers losses due to fire and flood;

- a *basic cyber insurance product* (denoted as *Cyber1*) that covers the risk of a data breach resulting in the loss of customer and employee data, including ensuing brand damage. It also covers the theft of confidential business information and intellectual property, including the loss of competitive advantage that accompanies this; and

- a more *comprehensive cyber insurance product* (denoted as *Cyber2*) that covers the same risks as the Cyber1 product as well as the risk of failure to deliver products and services. This includes the non-fulfilment of service and contractual agreements with respect to third parties (and the impact on the assets and business continuity of those third parties).

Median can select either the conventional insurance product, one of the cyber insurance products, or both the conventional insurance product and one of the cyber insurance products. It would not make sense for Median to choose more than one cyber insurance product, given their overlap in coverage.

We summarise the main features of each insurance product in Table 5.10. This includes the type of damage (i.e. the threat impacts on assets) that is covered and the percentage of the losses that are reimbursed. It also lists the price of each insurance product. As an example, the conventional insurance product covers 70% of the potential damage to IT systems and computer equipment and costs $3,000. Note that fines are not shown in the table since in Small Nation it is not legal for insurance to cover fines.

	Equip	MShare	Avail	CustEmp	BusInfo	Price
Conv	70%	–	–	–	–	3,000
Cyber1	–	50%	–	30%	30%	7,000
Cyber2	–	50%	50%	30%	30%	12,000

Table 5.10: Median's insurance product choices, according to coverage and prices

Attacker decision factors

We now look at the factors that adversaries take into account when deciding whether or not to attack. These include their uncertainties and objectives.

Uncertainties

In this case study the attackers have only one uncertainty, which we have called the *detection of the attacker* in Chapter 4 and elsewhere in this book. This refers to whether or not Median is able to identify the attacker who launched an attack against it.

Others using this model may wish to consider additional uncertainties, such as whether or not the attackers are able to obtain accurate information about the security controls of the organisation they want to attack. For the purposes of this case study, however, we assume that the attackers have been able to acquire such information.

Objectives

The attackers have a number of objectives to take into account as well. For the purposes of this case study, we have selected the five most relevant:

- *Maximising their gain in market share.* This is the extent of the increase in market share that an attacker can obtain from the company they attack (denoted as *Gain market share*). It is applicable in the case of an attack by a competitor.

- *Minimising their costs if detected.* These are the costs incurred by an attacker if he is identified (denoted as *Costs if detected*). In the case of an attack by a competitor, it might involve legal costs as well as reputational damage, loss of customers, etc. For a nation state, it could entail economic sanctions and military retaliation.

- *Maximising the customer and employee data obtained, including PII.* This is the amount of customer and employee data, including PII, that an attacker is able to steal (denoted as *Max customer & employee data*). Cybercriminal groups typically seek to sell this data on the black market for monetary gain. Competitors may do the same, or they may exploit the information themselves to gain competitive advantage. Nation state attackers typically capitalise on this information for strategic reasons.

- *Maximising the business information obtained.* This is the amount of business information that an attacker is able to steal (denoted as *Max business info*). The motivations of the actors interested in customer and employee data described above (cybercriminals, competitors, and nation states) are also applicable here. Some nation states may view business information as more valuable than customer and employee data. This is particularly the case for nation states whose policies are focused on commercial expansion.

- *Maximising the downtime for the attacked company.* This is the length of time for which a company that has been attacked is unable to provide services (denoted as *Max downtime*).

- *Minimising the implementation costs.* These are the financial costs to an attacker of carrying out an attack (denoted as *Implementation costs*).

Others using this model may want to consider additional objectives. For example, the national security value of the information held by a company is a significant decision factor for a nation state attacker.

Relevant objectives for each attacker

Not all of these objectives are relevant for all attackers. Table 5.11 shows which ones are relevant for which attackers in this case study. For example, Cybegangsta is especially interested in maximising its monetary gain and minimising its implementation and detection costs. Conversely, Modern Republic is not particularly concerned with implementation costs, as it has a dedicated team responsible for carrying out these types of activities that is funded from the government budget.

	Compeet	Antonymous	Cybegangsta	Modern Republic
Gain market share	x			
Costs if detected	x	x	x	x
Customer & employee data			x	
Business info			x	x
Downtime	x	x		x
Implementation costs	x	x	x	

Table 5.11: Objectives for each attacker

In Section 5.4, we show how we aggregate these objectives to determine the attackers' utility functions.

Constraints

We determine that the following constraints should be included in the model:

- *compliance with the Small Nation Cyber Essentials.* As discussed earlier, Small Nation recommends that a company like Median should include at least two security controls, (i) a firewall and internet gateway and (ii) an access control system, within its cybersecurity portfolio; and

- Median's annual *cybersecurity budget*, which is $72,000 for the coming year.

We designate the portfolio that complies with the Small Nation Cyber Essentials (a firewall and internet gateway and an access control system) as the *default portfolio*. These two security controls cost a combined $11,600, so once the default portfolio is implemented, the remaining cybersecurity budget is $72,000 − $11,600 = $60,400. It also means that the effective choice of security controls consists of (i) a sprinkler system, (ii) a flood door and flood risk management procedures, (iii) a DDoS mitigation system, (iv) a secure configuration, (v) a malware protection system, (vi) a patch vulnerability management system, and (vii) an intrusion detection system, since the default portfolio is already in place.

5.3   Model formulation

We present Median's CSRM decision-making process as a Multi-Agent Influence Diagram in Figure 5.1. We depict the five agents discussed in the previous section: Median (designated as the *Defender*) must decide on its cybersecurity resource allocation in order to protect itself against various threats, including those from its adversaries. These consist of Compeet, Antonymous, Cybegangsta, and Modern Republic (designated as the *Attackers*). To distinguish between the agents, nodes involving the Defender are white and nodes involving the Attackers are in different shades of grey; striped nodes are relevant to the decisions of the Defender and at least one of the Attackers.

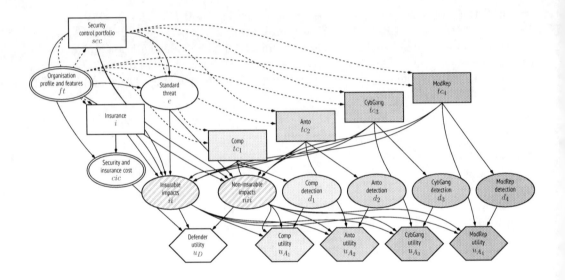

Figure 5.1: MAID for Median

The *organisation profile and features* (designated as ft) describe Median in terms of its profile, assets, and other features and is modelled as a deterministic node, which we thus represent as a double oval. In terms of the threats that could impact Median, to simplify the MAID we group the three types of unintentional threats (environmental, accidental, and non-targeted cyber) described in Section 5.2 under the banner of *standard threats* (designated as e). This is modelled as an uncertainty node and thus represented as an oval.

We detail the targeted cyber threats intentionally directed at Median as: *targeted cyber threats from Compeet* (designated as tc_1), *targeted cyber threats from Antonymous* (designated as tc_2), *targeted cyber threats from Cybergangsta* (designated as tc_3), and *targeted cyber threats from Modern Republic* (designated as tc_4). They are all modelled as decision nodes (since they involve decisions made by the Attackers) and thus represented as rectangles.

To further simplify the MAID, we also group the threat impacts on assets described in Section 5.2 into two broad categories: *insurable impacts* (designated as ii) that are partially covered by insurance and *non-insurable impacts* (designated as nii) that are not covered by insurance. These are modelled as uncertainty nodes so are represented as ovals. They are striped because they are relevant to the decisions of the Defender as well as the Attackers.

To mitigate the likelihood and/or impacts of the threats, Median must determine how to allocate its resources between: its *security control portfolio* (designated as *sec*), which consists of various security control products to protect and defend against cyber attacks, and *insurance* (designated as *i*), which consists of one or more insurance products that will compensate it in the event of an attack. They are both modelled as decision nodes (since they involve decisions made by Median) and thus depicted as rectangles.

These instruments entail *security and insurance costs* (designated as *cic*), which is a deterministic node and hence denoted as a double oval. They must also take certain constraints into account, including the security controls that are needed to comply with the Small Nation Cyber Essentials as well as Median's cybersecurity budget, as discussed in Section 5.2.

Having identified all of the elements that are relevant for Median, we can now build Median's preference model, which is determined by its utility function. We therefore model the *Defender's expected utility* (designated as u_D) as a value node and thus depict it as a hexagon. This utility function covers risk attitude aspects as well.

In terms of the elements that are relevant for the Attackers, a key factor is whether or not Median is able to identify who it has been attacked by. We depict this as: *detection of Compeet as the attacker* (designated as d_1), *detection of Antonymous as the attacker* (designated as d_2), *detection of Cybergansta as the attacker* (designated as d_3), and *detection of Modern Republic as the attacker* (designated as d_4). These are all modelled as uncertainty nodes and hence depicted as ovals.

Having identified all of the elements that are relevant for the Attackers, we can now build the Attackers' preference models, which are also based on their utility functions. We therefore model *Compeet's expected utility* (designated as u_{A_1}), *Antonymous's expected utility* (designated as u_{A_2}), *Cybegansta's expected utility* (designated as u_{A_3}), and *Modern Republic's expected utility* (designated as u_{A_4}) as value nodes and, accordingly, depict them as hexagons.

This case study illustrates how the basic CSRM model and the other models that we presented in Chapter 4 can serve as templates that can be expanded, reduced, or otherwise modified. In this case study, we have added two additional Attackers compared to the basic CSRM template model in Chapter 4. Conversely, we have not included recovery controls, as the company in this case study is not considering purchasing them.

5.4 Model components

We now show how to assess the various components of the model. Section 5.4.1 looks at non-targeted threats, modelling their likelihood of occurring and their impacts on Median (depending on Median's choice of security controls). It also considers whether the impacts are covered by any of the insurance products. In Section 5.4.2, we turn to targeted threats. After developing a general utility model for Attackers, we consider the types of attacks they carry out, their impacts (depending on Median's choice of security controls), and the Attacker's probability of being identified as the perpetrator, enabling us to develop a specific utility model for each Attacker. We use this to run a simulation to determine the Attacker's likelihood of carrying out an attack against Median. Section 5.4.3 summarises the impacts of each threat on Median, evaluating their costs for Median. We also consider other relevant costs and gains, using this to develop Median's utility model.

This section builds upon the work of Rios Insua et al. (2019) and Couce-Vieira et al. (2020b), who propose ways to assess some of these components. Therefore, we only provide detailed explanations for the elements that are new in this case study. For the others, we give a more condensed explanation and denote these with a ‡ symbol, so that others wishing to apply this model can refer back to their work for the full description.

5.4.1   Likelihood and impacts of non-targeted threats

For each non-targeted threat, we demonstrate how we can model its likelihood and its impacts. We then discuss whether the impacts can be covered by insurance.

Environmental threats

Fire ‡

Likelihood

To model the number of fires that an industrial installation such as Median can expect in a given year, we start by estimating the probability of fire to be 2.2%, or 0.022, annually, based on data analysed in Rios Insua et al. (2019). We can then model this using a Poisson $\mathcal{P}(0.022)$ distribution.[1] This tells us that the probability of Median experiencing no fires in a year is roughly 98% and the probability of it experiencing more than one fire in a year is 0.02%, among other things.

Impacts

A fire will impact Median's IT systems and computer equipment, as indicated in Table 5.8. The impact depends on the percentage of facility degradation caused by the fire. Assuming that the IT systems and computer equipment are evenly distributed throughout the premises, a fire lasting 120 minutes would destroy it completely.

   The implementation of a sprinkler system as a security control can reduce the impact of the fire, as indicated in Table 5.9. The fire's duration depends on whether or not Median has implemented the sprinkler system. If it has, the duration of the fire can be described using a Triangular $\mathcal{T}ri(0.8, 63, 10)$ distribution (with the fire duration measured in minutes). Otherwise, it is described with a Gamma $\Gamma(0.85, 0.011)$ distribution. Details on how these distributions were derived are available in Rios Insua et al. (2019). The impact of a fire on IT systems and computer equipment is insurable through the conventional insurance product, as shown in Table 5.10.

Flood

Likelihood

To model the number of floods that Median can expect in a given year, we first determine that flooding is about 20 times more likely than fire in industrial installations,[2] according to data from Estamos Seguros (2017). Therefore, we assume that the number of annual floods that Median could experience follows a Poisson $\mathcal{P}(0.44)$ distribution (obtained by calculating 0.022×20).

---

[1]We employ the parametric notation used in R when describing a distribution.

[2]Flooding is a multiple source problem, meaning it could be the result of a number of causes such as heavy rains, broken water lines, or frozen pipes.

Impacts

A flood will impact Median's IT systems and computer equipment, as indicated in Table 5.8. The implementation of flood doors and flood risk management procedures as a security control will help reduce the flood's impact, as shown in Table 5.9. If Median implements them, industry estimates suggest that this reduces the likelihood of the flood causing damage to Median by a factor of 10. We model this with a Poisson $\mathcal{P}(0.044)$ distribution (obtained by calculating $0.44 \div 10$).

When flood risk management procedures are in place to facilitate a rapid response to a flood, according to Tech!Espresso (2020) in some 95% of cases the damage to IT systems and computer equipment can be repaired, at an estimated cost of $200 per device. Without such procedures, the damage can only be repaired in 70% of cases. In the remaining cases (5% if flood risk management procedures are implemented and 30% if they are not), the devices need to be replaced. We assume that all devices are affected by the flood. The impact of a flood on IT systems and computer equipment is insurable through the conventional insurance product, as shown in Table 5.10.

Accidental threats

Employee error ‡

Likelihood

To model the number of employee errors $a_{emperror}$ in a given year, for a company of Median's size we start by estimating the probability of an employee error to be 6%, or 0.06, per year. We then represent this with a Poisson $\mathcal{P}(0.06)$ distribution, according to Couce-Vieira et al. (2020b).

As shown in Table 5.9, the implementation of an access control system as a security control reduces the likelihood of an employee error occurring. We estimate that it will lower the probability by 50%. We can take this into account through a risk reduction coefficient *red*, as shown in Couce-Vieira et al. (2020b), which considers that the implementation of the security control can have an impact on 66% of the risk. We describe this as

$$red = 0.34 + 0.66 \times (1 - 0.50 sec_{access}),$$

where $sec_{access} = 1$ if the security control is implemented and 0 otherwise.[3] (The constant 0.34 is obtained by calculating $1 - 0.66$.)

This means that the number of employee errors per year is actually modelled as a Poisson $\mathcal{P}(0.06 \times red)$ distribution. Since Median has implemented an access control system as part of the Small Nation Cyber Essentials scheme, $sec_{access} = 1$. Therefore, $red = 0.67$, giving us a Poisson $\mathcal{P}(0.0402)$ distribution (obtained by calculating 0.06×0.67).

Impacts

The potential impacts of an employee error involve (i) the exposure or loss of customer and employee data, (ii) the exposure or loss of business information, as well as (iii) fines, as shown in Table 5.8. For customer and employee data as well as business information, we can model the impact of the exposure or loss of these records as

$$nii_{recexp|a_{emperror}} = 0.1243 \times y_{recexp} \times a_{emperror},$$

where y_{recexp} follows a Uniform $\mathcal{U}(0, ft_{records})$ distribution, if $a_{emperror} > 0$. See Couce-Vieira et al. (2020b) for an explanation of how this is derived. In the case of customer and

---

[3]In general, we denote $sec_x = 1$ if the control x is implemented and $sec_x = 0$ otherwise.

employee data, the number of records $ft_{records}$ is 200,000 and the value of each record is estimated at \$825. In the case of business information, the number of records $ft_{records}$ is 66,000 and the value of each record is estimated at \$3,000. An employee error that results in the exposure or loss of these records can be covered by either of the cyber insurance products, as shown in Table 5.10.

In terms of fines, it is estimated that the data protection regulator will fine Median approximately \$60,000 in a situation in which more than 11,000 records are exposed, according to Valls (2019). These fines are not insurable, however, since as previously mentioned Small Nation does not allow for insurance to cover fines.

Misconfiguration

Likelihood

To model the number $a_{misconf}$ of misconfigurations that Median is likely to experience in a given year, we start with the estimate that there is a 10%, or 0.10, yearly probability of this occurring, based on data from GlobalDots (2019). We can represent this with a Poisson $\mathcal{P}(0.10)$ distribution.

From Table 5.9, we see that the implementation of an access control system and/or a secure configuration as security controls can reduce the likelihood of a misconfiguration. We estimate that:

- Since an access control system has already been implemented as part of the Small Nation Cyber Essentials, this reduces the probability of a misconfiguration by 60%.

- If a secure configuration is implemented, this reduces the probability of a misconfiguration by 50%.

This means that we can calculate the yearly probability of a misconfiguration as

$$0.04 \times (1 - 0.5sec_{secconf}).$$

(We obtain 0.04 by calculating 60% of 0.10.)

We can therefore model the number of misconfigurations per year using a Poisson $\mathcal{P}(0.04 \times (1 - 0.5sec_{secconf}))$ distribution, where $sec_{secconf} = 1$ if the security control is implemented and 0 otherwise.

Impacts

The potential impacts of a misconfiguration might involve (i) damage to IT systems and computer equipment, (ii) the exposure or loss of customer and employee data, and (iii) the exposure or loss of business information, as reflected in Table 5.8. In terms of IT systems and computer equipment, the probability that a server is damaged due to a misconfiguration is 0.005, according to Reynolds (2019). We assume that the cost of repairing the damage associated with each misconfiguration is the same as the replacement cost.

From Table 5.9, the implementation of an access control system and/or secure configuration as security controls can reduce the impact of a misconfiguration as well. We can also estimate that:

- Since an access control system has been implemented, this reduces the probability of a misconfiguration resulting in damage by 60%.

- If a secure configuration is implemented, this reduces the probability of a misconfiguration resulting in damage by 50%.

This means that we can calculate the probability of a misconfiguration resulting in damage occurring as

$$0.002 \times (1 - 0.5sec_{secconf}).$$

(We obtain 0.002 by calculating 60% of 0.005.) This damage to IT systems and computer equipment is not insurable, as shown in Table 5.10.

In terms of customer and employee data as well as business information, we can model these as we did in the preceding section on employee error. That is, we can model the impact of the exposure or loss of these records as

$$nii_{recexp|a_{misconf}} = 0.1243 \times y_{recexp} \times a_{misconf},$$

where y_{recexp} follows a Uniform $\mathcal{U}(0, ft_{records})$ distribution, if $a_{misconf} > 0$. Once again, in the case of customer and employee data, the number of records $ft_{records}$ is 200,000 and the value of each record is \$825; in the case of business information, the number of records $ft_{records}$ is 66,000 and the value of each is \$3,000. The exposure or loss of both of these types of records are insurable through either of the cyber insurance products, as shown in Table 5.10.

Non-targeted cyber threats

Ransomware

Likelihood

To model the number $a_{rsmware}$ of ransomware attacks on Median in a given year, we begin with the estimate that Median has a 5.28%, or 0.0528, yearly probability of experiencing such an attack. We represent this with a Poisson $\mathcal{P}(0.0528)$ distribution, based on data analysed in Couce-Vieira et al. (2020b).

From Table 5.9, we see that the implementation of various security controls—a firewall and internet gateway, a secure configuration, a malware protection system, a patch vulnerability management system, and/or an intrusion detection system—can reduce the likelihood of a ransomware attack occurring. We estimate that:

- Since a firewall and internet gateway has already been implemented as part of the Small Nation Cyber Essentials, this reduces the probability of a ransomware attack by 50%.

- If a secure configuration is implemented, this reduces the probability of a ransomware attack by 50%.

- If a malware protection system is installed, this reduces the probability of a ransomware attack by 90%.

- If a patch vulnerability management system is integrated, this reduces the probability of a ransomware attack by 50%.

- If an intrusion detection system is implemented, this reduces the probability of a ransomware attack by 13%.

We can take this into account through a risk reduction coefficient red, as shown in Couce-Vieira et al. (2020b), which considers that the implementation of various security controls can have an influence on 66% of the risk. We describe this as

$$red = 0.34 + 0.66(1 - 0.5sec_{firewall})(1 - 0.5sec_{secconf})(1 - 0.9sec_{mps}) \times$$
$$\times (1 - 0.5sec_{pvm})(1 - 0.13sec_{ids}).$$

Given that Median has already implemented the firewall and internet gateway, $sec_{firewall} = 1$, so that

$$red = 0.34 + 0.33(1 - 0.5sec_{secconf})(1 - 0.9sec_{mps})(1 - 0.5sec_{pvm})(1 - 0.13sec_{ids}).$$

This therefore enables us to model the annual number of ransomware attacks using a Poisson $\mathcal{P}(0.0528 \times red)$ distribution.

Impacts

As indicated in Table 5.8, the main impacts of a ransomware attack are (i) loss of availability, i.e. inability to deliver services, (ii) the exposure or loss of customer and employee data, and (iii) the exposure or loss of business information. We estimate that the loss of availability due to a ransomware attack follows a Gamma $\Gamma(10, 1)$ distribution (in which loss of availability is measured in hours). This is based on expert judgement; we relied on the assessments of three experts regarding the duration quantiles for probabilities of 0.1, 0.5, and 0.9. We averaged their responses, which gave us the values of 6.2, 9.6, and 14.2 hours. Using the first and last values, we estimated the parameters at approximately 10 and 1, and used the median as a consistency check, which confirmed that this gives us a reasonably good approximation. We did not use expert calibration as in Torres et al. (2020). The loss of availability as a result of a ransomware attack is covered by the more comprehensive cyber insurance product, or Cyber2, as per Table 5.10.

Regarding customer and employee data as well as business information, we can use the same approach used in the employee error section. However, we use a coefficient of 0.4 instead of 0.1234, that is

$$nii_{recexp|a_{rsmware}} = 0.4 \times y_{recexp} \times a_{rsmware},$$

where y_{recexp} follows a Uniform $\mathcal{U}(0, ft_{records})$ distribution, if $a_{rsmware} > 0$. A detailed explanation of how the coefficient 0.4 is derived is available in Couce-Vieira et al. (2020b). Again, in the case of customer and employee data, $ft_{records}$ is 200,000 and the value of each record is \$825; in the case of business information, $ft_{records}$ is 66,000 and each record is \$3,000. The exposure or loss of both of these types of records are insurable through either of the cyber insurance products, as per Table 5.10.

Virus ‡

Likelihood

To model the number a_{virus} of virus infections that Median is likely to experience in a year, we start with monthly data that we have been able to obtain. According to Rios Insua et al. (2019), a baseline estimate for the probability of a computer being infected in a given month, q_1, is 33%, or 0.33. The number of computers infected by a virus in a given month can be modelled using a Binomial $\mathcal{B}(h_1, q_1)$ distribution, with h_1 being the number of computers. A baseline estimate for the probability of a server being infected in a given month, q_2, is 23%, or 0.23. The number of servers infected by a virus in a given month can be modelled using a similar Binomial $\mathcal{B}(h_2, q_2)$ distribution, with h_2 being the number of servers.

From Table 5.9, we see that the implementation of a number of security controls—a firewall and internet gateway, a secure configuration, a malware protection system, a patch vulnerability management system, and/or an intrusion detection system—can reduce the likelihood of a virus infection occurring. We estimate that:

- Since Median has already implemented a firewall and internet gateway as part of the Small Nation Cyber Essentials, the probability of a computer being infected is thought to only be approximately 0.5%, or 0.005. The probability of a server being infected is estimated to be even lower, at 0.05%, or 0.0005. This significantly reduces the threat. These estimates are based on expert judgement, as described in Rios Insua et al. (2019).

- If a secure configuration is implemented, this reduces the probability of infection by 25%.
- If a malware protection system is installed, this reduces the probability of infection by 75%.
- If a patch vulnerability management system is integrated, this reduces the probability of infection by 50%.
- If an intrusion detection system is implemented, this reduces the probability of infection by 50%.

Therefore, the probability of a computer being infected by a virus in a given month can be expressed as

$$q_1 = \left[0.005sec_{firewall} + 0.33(1 - sec_{firewall})\right](1 - 0.25sec_{secconf}) \times$$
$$\times (1 - 0.75sec_{malprot})(1 - 0.5sec_{pvm})(1 - 0.5sec_{ids}).$$

Since Median has implemented a firewall and internet gateway, $sec_{firewall} = 1$, so that

$$q_1 = 0.005 \times (1 - 0.25sec_{secconf})(1 - 0.75sec_{malprot})(1 - 0.5sec_{pvm})(1 - 0.5sec_{ids}).$$

Similarly, the probability of a server being infected by a virus in a given month can be expressed as

$$q_2 = \left[0.0005sec_{firewall} + 0.33(1 - sec_{firewall})\right](1 - 0.25sec_{secconf}) \times$$
$$\times (1 - 0.75sec_{malprot})(1 - 0.5sec_{pvm})(1 - 0.5sec_{ids}).$$

Again, since Median has implemented a firewall and internet gateway, $sec_{firewall} = 1$, so that

$$q_2 = 0.0005 \times (1 - 0.25sec_{secconf})(1 - 0.75sec_{malprot})(1 - 0.5sec_{pvm})(1 - 0.5sec_{ids}).$$

Since we are interested in the likelihood of being infected by a virus over the coming 12 months, we multiply $q_1 \times 12$ and $q_2 \times 12$ to obtain yearly probabilities. The number of computers h_1 that Median has is 300 and the number of servers h_2 is 40. Therefore, we can model the number of virus infections in a given year using the Binomial $\mathcal{B}(300, q_1 \times 12)$ distribution for computers and using the the Binomial $\mathcal{B}(40, q_2 \times 12)$ distribution for servers.

Impacts

As Table 5.8 shows, the main impacts of a virus are (i) damage to IT systems and computer equipment, (ii) loss of availability, i.e. inability to deliver services, (iii) the exposure or loss of customer and employee data, and (iv) the exposure or loss of business information. To repair each computer or server infected by a virus, we estimate the cost to be $31, or the price of two technician hours. These costs are covered by both cyber insurance options, as in Table 5.10.

In terms of loss of availability, one way to model this is to assess Median's total productivity loss s. We estimate that when a computer is infected, its usability will be impacted for approximately 28 hours and model the amount of time lost w per employee with a Uniform $\mathcal{U}(0, 0.05)$ distribution, as in Rios Insua et al. (2019). Given that the average hourly wage of an employee is \$20 per hour and we recall that Median has 50 employees, we can estimate Median's productivity loss using

$$s = 28 \times w \times 20 \times 50.$$

This productivity loss is not covered by any of the insurance products.

In terms of customer and employee data and business information, we can again use the same approach employed in the employee error section. That is, we can model the impact of the exposure or loss of these records as

$$nii_{recexp|a_{virus}} = 0.1243 \times y_{recexp} \times a_{virus},$$

where y_{recexp} follows a Uniform $\mathcal{U}(0, ft_{records})$ distribution, if $a_{virus} > 0$. Once again, in the case of customer and employee data, $ft_{records}$ is 200,000 and the value of each record is \$825; in the case of business information, $ft_{records}$ is 66,000 and the value of each is \$3,000. The exposure or loss of these records is covered by both of the cyber insurance products, as shown in Table 5.10.

5.4.2   Likelihood and impacts of targeted cyber threats

We now turn to targeted cyber threats. First, we describe a general utility model for Attackers. Next, we examine each Attacker individually. For each, we consider the type(s) of attacks they engage in and the impacts. Then we determine their probability of being identified if they carry out an attack, i.e. their detection probability. We use this to build a specific utility model for each Attacker. This enables us to better understand each Attacker's strategic thinking when it comes to attacking Median. We then determine their likelihood of carrying out an attack, i.e. their attack probability. This is done by running a simulation to assess their optimal attack strategy and forecast their expected behaviour, depending on Median's choices of security controls.

General preference model for Attackers

We build a generic model for the preferences of Attackers (which is specified through their utility functions). This serves for all four Attackers and is based on their objectives presented in Table 5.11. In order to be able to determine the utilities of the Attackers, we must first translate these objectives into quantifiable monetary terms.

- The Attacker objective of maximising their gain in market share can be converted into what we call *gain in market share*, or the monetary value of the increase in market share that the Attacker can obtain by attacking Median (denoted as m).

- The Attacker objective of minimising their costs if detected can be converted into the *costs if detected*, or the monetary value of the expected costs to the Attacker if he is identified (denoted as c_d).

- The Attacker objective of maximising the customer and employee data obtained, including PII, can be converted into the *value derived from customer and employee data, including PII*, or the value expressed in monetary terms that the Attacker can derive by either selling one unit of customer and employee data, including PII, or by employing it for his own use (denoted as v_r).

- The Attacker objective of maximising the business information obtained can be converted into the *value derived from business information*, or the value expressed in monetary terms that the Attacker can derive by either selling one unit of business information or by employing it for his own use (denoted as v_b).

- The Attacker objective of maximising the downtime of the company they attack can be converted into the *value derived from downtime*, or the value expressed in monetary terms that the Attacker can derive per unit of downtime that he causes to Median (denoted as v_d).

- The Attacker objective of minimising the implementation costs can be converted into the *implementation costs*, or the financial costs to the Attacker of carrying out the attack (denoted as c_i).

These are summarised in Table 5.12.

Objectives in monetary terms	Variable
Gain in market share	m
Costs if detected	c_d
Value from customer & employee data	v_r
Value from business info	v_b
Value from downtime	v_d
Implementation costs	c_i

Table 5.12: Attacker objectives in monetary terms

Given that three of the above objectives are measured in units, we also need to define variables to capture the total number of units. We therefore create variables to express the *number of customer and employee data records, including PII* (denoted as n_r), the *number of business records* (denoted as n_b), and the *number of downtime units* (denoted as d). These are summarised in Table 5.13.

Objectives in total units	Variable
Customer & employee data units	n_r
Business info units	n_b
Downtime units	d

Table 5.13: Attacker objectives in total units

We then aggregate these objectives, now that they have been monetised, using

$$c = \lambda_1 m - \lambda_2 c_d + (n_r \times v_r) + (n_b \times v_b) + (d \times v_d) - \lambda_3 c_i,$$

where c is an Attacker's earnings. Not all objectives are relevant for all Attackers. Therefore, we also define dummy weights λ_i, $i = 1, 2, 3$ such that $\lambda_i = 1$ if the corresponding objective is relevant to an Attacker and 0 otherwise.

Based on the Attacker's earnings c, we adopt a risk-prone utility function

$$u_A(c) = a_A \times \exp(\rho_A \times c) + b_A,$$

where the constants $a_A > 0$, b_A, and $\rho_A > 0$ are used to scale the utility function to the interval $[0, 1]$. However, since we do not have sufficient information about the Attacker to properly elicit $u_A(c)$, we model the Defender's beliefs about the Attacker's parameters using the distributions of a_A, b_A, and ρ_A, as detailed in the Appendix and discussed in Section 4.3.4. Moreover, whenever relevant, we need to model the uncertainty surrounding the parameters v_r, v_b, and v_d. These uncertainties serve to provide the random utility function of an Attacker.

Specific utility models and attack probabilities for each Attacker

Compeet

We assume that Compeet is primarily interested in carrying out DDoS attacks, as reflected in Table 5.7. Compeet's main objectives consist of (i) increasing his market share, (ii) minimising his implementation costs of carrying out an attack, and (iii) minimising his costs if detected (i.e. if he is identified as having carried out the attack), as shown in Tables 5.11 and 5.12. Note that while increasing his market share is closely intertwined with maximising the downtime of his main competitor Median, the latter is not necessarily a primary objective for Compeet.

DDoS attack ‡

In order to determine the duration l in hours of a successful DDoS attack, we must consider the intensity and sophistication of the attack as well as the security controls implemented. As shown in Table 5.9, the primary security control used to defend against a DDoS attack is a DDoS mitigation system.

We begin with an estimate that standard attacks last 4 hours, averaging 1 gbps with peaks of up to 10 gbps (Verisign, 2017). Therefore, we start by modelling l_j, the length of the jth individual DDoS attack, using a Gamma $\Gamma(4,1)$ distribution. We then assume that Compeet hires a third party to carry out the attacks that is capable of attacking at approximately 5 gbps, which we thus represent with a Gamma $\Gamma(5,1)$ distribution.

To take into account the uncertainty around an attack's average duration and its dispersion, we actually model l_j as a random Gamma $\Gamma_{\text{length}}(v, v/\mu)$ distribution with v following a Uniform $\mathcal{U}(3.6, 4.8)$ distribution and v/μ following a Uniform $\mathcal{U}(0.8, 1.2)$ distribution, so that its average duration could be between 3 and 6 hours.

We then subtract the traffic s_{gbps} that the DDoS mitigation system would divert to determine whether or not a DDoS attack is likely to be successful. Given that the DDoS mitigation system that Median is considering is able to defend against attacks of up to 10 gbps, we can model

$$l = \sum_{j}^{a} l_j,$$

where a is the number of days (if any) for which Compeet decides to carry out the attack campaign, with $l_j \sim \Gamma_{\text{length}}$ if $\Gamma_{\text{gbps}} - 10 > 0$ and $l_j = 0$ otherwise. On the other hand, if the DDoS mitigation system is not implemented, $l_j \sim \Gamma_{\text{length}}$.

Impacts and outcomes ‡

The attack impacts and outcomes for Compeet include (i) his gain in market share m and (ii) his implementation costs c_i. Compeet's gain in market share m is closely linked to the amount of downtime that Median experiences. To assess this, we assume that Median loses her market share at a linear rate until it has fallen to close to 0 (which for ease of modelling we take to be 0) after approximately 5-8 days of unavailability.

Then, to determine the fastest possible rate at which Median could lose market share, we recall that Median's market share is 50%, or 0.5. Given that 5 days × 24 hours per day is 120 hours, we calculate $0.5 \div 120$ to obtain a loss in market share per hour of 0.0042. To calculate the slowest possible rate at which she could lose market share, given that 8 days × 24 hours per day is 192 hours, we compute $0.5 \div 192$ to give us a loss in market share per hour of 0.0026.

We can capture this uncertainty surrounding the rate of loss of market share by modelling the loss rate R as a Uniform $\mathcal{U}(0.0026, 0.0042)$ distribution. Then, the distribution describing the loss in market share is 15,000,000 min$(0.5, l \times R)$, given that the total value of the market is \$15,000,000. Since Compeet is Median's primary competitor, for ease of modelling we assume that the market share lost by Median is equivalent to Compeet's gain in market share m.

As to the implementation costs c_i that Compeet incurs, we estimate that launching a DDoS attack would cost approximately \$33 per hour, or \$792 per day, according to Imperva (2019). Therefore, for a number of attack days a the implementation costs are $c_i = 792 \times a$.

Detection costs and probability ‡

We now examine Compeet's expected costs if detected c_d and his probability of detection (i.e. of being identified as the perpetrator). If Compeet is identified as being the Attacker, the costs that he faces involve serious reputational damage resulting in the loss of customers as well as major legal costs, as mentioned earlier. This could include having to pay compensation to Median. To capture these uncertainties surrounding the costs, we use expert judgement to model the expected costs c_d that Compeet faces if detected using a Normal distribution with \$2,430,000 as the mean and a standard deviation of \$400,000. This is therefore represented as a Normal $\mathcal{N}(2{,}430{,}000; 400{,}000)$ distribution if the attack is detected and 0 otherwise.

To determine Compeet's probability of detection over the course of his entire attack campaign, we use expert judgement to estimate this to be 0.2%, or 0.002, for each individual DDoS attack that Compeet launches against Median. Given the uncertainty associated with this, we model it using a Beta $\mathcal{B}e(2, 998)$ distribution (which has a mean of 0.002). Since Compeet needs to decide as to the total number of days a for which to carry out the attack campaign, we model his probability of detection over the course of the entire attack campaign using a random Binomial $\mathcal{B}(a, 0.002)$ distribution which outputs "detected" if $\mathcal{B}(a, \phi) > 0$ with $\phi \sim \mathcal{B}e(2, 998)$ and "not detected" otherwise.

Compeet's utility

We can now build a preference model for Compeet (that is, his random utility). Given the objectives that are relevant for him, Compeet's earnings c (i.e. the objectives that he has achieved, calculated in monetary terms) depend on his gain in market share m, his implementation costs c_i, and his expected costs if detected c_d, thus giving

$$c = m - c_i - c_d.$$

Then, we assess the distributions of a_{COMP}, b_{COMP}, and ρ_{COMP}, as shown in the Appendix. For this, we use a lower bound estimate for Compeet's earnings of $c_* = -3{,}100{,}000$ and an upper bound estimate of $c^* = 6{,}100{,}000$ and assess that $u(0) \sim \mathcal{U}(0.1, 0.3)$.

Attack probabilities

We now examine Compeet's strategic thinking when it comes to attacking Median. More specifically, Compeet must decide whether or not to attack Median, and if so, on the number a of attack days for which to wage the campaign. To do this, we carry out a simulation in order to estimate the distribution of optimal attacks for Compeet to conduct. For each possible cybersecurity portfolio, we simulate the attack process 1,000 times with just Compeet as the Attacker. This makes it possible to forecast Compeet's behaviour, depending on Median's choices regarding security controls. As shown in Table 5.9, the only effective security control against a DDoS attack is a DDoS mitigation system, so we show two instances: one in which Median's security control portfolio includes a DDoS mitigation system and one in which it does not. We present the attack probabilities that result from the simulation in Figure 5.2. The options considered in the simulation range from Compeet not attacking Median at all (i.e. attacking it for 0 days) to Compeet attacking Median for anywhere from 1 to 30 days.

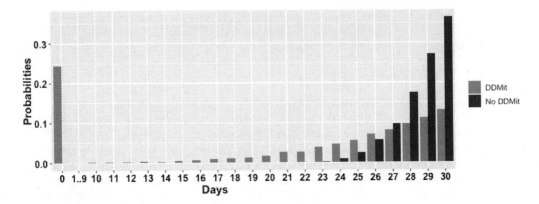

Figure 5.2: Compeet's probabilities of attacking Median for 0 to 30 days, depending on whether Median has a DDoS mitigation system

We can see that the inclusion of a DDoS mitigation system in the security control portfolio has a significant deterrent effect on Compeet. For one, it considerably reduces the probability that Compeet will attack Median. Furthermore, it forces Compeet to expend significantly more effort and carry out a prolonged attack campaign in order to have a reasonable chance of success. Conversely, the absence of a DDoS mitigation system in the security control portfolio means that Compeet is highly likely to attack Median and engage in an even more intense, long-lasting attack campaign.

Antonymous

Antonymous is highly adept at carrying out DDoS attacks, which are his weapon of choice, as indicated in Table 5.7. His main objectives are to (i) maximise the downtime for Median, (ii) minimise his implementation costs, and (iii) minimise his costs if he is detected as being the Attacker, as shown in Tables 5.11 and 5.12.

DDoS attack

We can apply the same approach as for Compeet in order to determine the duration l of a successful DDoS attack. The only difference is that Antonymous can carry out the attack on his own, instead of needing to hire a third party to do so.

Impacts and outcomes

The attack impacts and outcomes for Antonymous include (i) the value that it derives from the downtime that Median experiences v_d and (ii) the implementation costs c_i of the attack. The duration of the downtime experienced by Median is equivalent to the duration l of a successful DDoS attack, which, as explained above, can be obtained in the same manner as for Compeet. We can thus model Median's downtime duration as

$$l = \sum_j^a l_j,$$

where $l_j \sim \Gamma_{\text{length}}$ if $\Gamma_{\text{gbps}} - 10 > 0$ and $l_j = 0$ otherwise.

To determine Antonymous's value per unit of downtime v_d, we start with a reference value of $10,000. However, Antonymous has an important tradeoff to consider when it comes to the length of the downtime that he causes Median, since this also increases his probability of being detected and his expected costs if detected c_d. To capture the uncertainties surrounding this, we can represent Antonymous's value per unit of downtime v_d using a Uniform $\mathcal{U}(0.8 \times 10,000, 1.2 \times 10,000)$ distribution.

As to the implementation costs c_i, we estimate that launching a DDoS attack would cost Antonymous approximately $600 per day. Therefore, for a number of attack days a the implementation costs are $c_i = 600 \times a$.

Detection costs and probability

We now examine Antonymous's expected costs if detected c_d and his probability of detection $hd_{probability}$ (i.e. his probability of being identified as the perpetrator). We can model Antonymous' expected costs if detected c_d using

$$c_d = hd_{probability} \times hd_{cost},$$

where hd_{costs} are the costs that Antonymous faces if detected.

When modelling these factors, we must take into account the tradeoff described above with the value per unit of downtime v_d, given that increasing the length of the downtime that Antonymous causes Median also increases his probability of being detected and his expected costs if detected c_d. To capture the uncertainties surrounding this, we model Antonymous's probability of detection $hd_{probability}$ in the same manner as for Compeet, by making use of a Binomial $\mathcal{B}(a, 0.002)$ distribution. We can use a similar approach to model hd_{costs}, using expert judgement to determine the lowest and highest possible costs to Antonymous ($300,000 and $450,000 respectively), then applying a Uniform distribution. We represent this as a Uniform $\mathcal{U}(300,000; 450,000)$ distribution.

Antonymous's utility

We can now build a preference model for Antonymous (that is, his random utility). Given the objectives that are relevant, Antonymous's earnings c (i.e. the objectives that he has achieved, calculated in monetary terms) depend on the downtime that he has caused Median (measured in terms of Antonymous's monetary gain per unit of downtime v_d multiplied by the number of units of downtime d), his implementation costs c_i, and his expected costs if detected c_d, thus giving

$$c = v_d \times d - c_i - c_d.$$

We then obtain the distributions of a_{ANT}, ρ_{ANT}, and b_{ANT}, as shown in the Appendix, using a lower bound estimate for Compeet's earnings of $c_* = -600,000$, and upper bound estimate of $c^* = 1,200,000$, and assessing that $u(0) \sim \mathcal{U}(0.05, 0.1)$.

Attack probabilities

We now turn to Antonymous's strategic thinking when it comes to attacking Median. Notably, Antonymous must decide whether or not to attack Median, and if so, on the number of attack days a for which to wage the campaign. As with Compeet, we carry out a simulation in order to estimate the distribution of optimal attacks for Antonymous to conduct. For each possible cybersecurity portfolio, we simulate the attack process 1,000 times with just Antonymous as the Attacker. This makes it possible to forecast Antonymous's behaviour, depending on Median's choices regarding security controls. Again, since the only effective security control against a DDoS attack is a DDoS mitigation system, we only show two instances: one in which Median's security control portfolio includes a DDoS mitigation system and one in which it does not. We present the attack probabilities generated by the simulation in Figure 5.3. As with Compeet, the options considered in the simulation range from not attacking Median at all to attacking it for anywhere from 1 to 30 days.

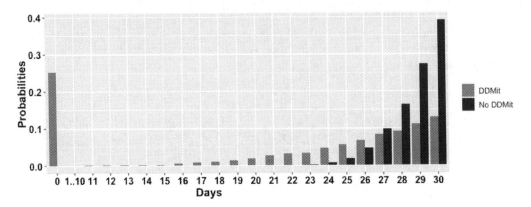

Figure 5.3: Antonymous's probabilities of attacking Median for 0 to 30 days, depending on whether Median has a DDoS mitigation system

The results for Antonymous are similar to those of Compeet. We can see that the inclusion of a DDoS mitigation system in the security control portfolio has a significant deterrent effect on Compeet. For one, it considerably reduces the probability that Compeet will attack Median. In addition, it forces Compeet to expend significantly more effort and carry out a prolonged attack campaign if it is to have any chance of success. Conversely, the lack of a DDoS mitigation system in the security control portfolio means that Compeet is much more likely to attack Median and to engage in an intense, long-lasting attack campaign.

Cybegangsta

Cybegangsta mainly engages in social engineering attacks, as shown in Table 5.7. His primary objectives are to maximise his monetary gain, either (i) from selling stolen customer and employee data or (ii) from selling stolen business information; (iii) to minimise his implementation costs; and (iv) to minimise his costs if he is detected as being behind the attack, as shown in Tables 5.11 and 5.12.

Social engineering attack

To model the number a_{soceng} of social engineering attacks in a given year, we use a similar approach to the one we used to model employee error. The two cases are analogous, the main difference being that in a social engineering attack an agent (Cybegangsta in this case) intentionally tries to trick employees into performing IT actions that will harm the company.

The implementation of various security controls—a firewall and internet gateway, an access control system, and/or an intrusion detection system—can reduce the likelihood of a company's employees being fooled by a social engineering attack, as shown in Table 5.9. More specifically:

- Since the default portfolio consisting of a firewall and internet gateway and an access control system has already been implemented as part of the Small Nation Cyber Essentials, the probability of Median's employees falling for a social engineering attack is expected to be just 0.9%, or 0.009, in a given year, based on comparable estimates for a company of Median's size from Positive Technologies (2018).

- If an intrusion detection system is implemented, this reduces the probability of Median's employees falling for a social engineering attack even further, to 0.5%, or 0.005.

Impacts and outcomes ‡

The attack impacts for Cybegangsta involve (i) his monetary gain from selling stolen customer and employee data, v_r (ii) his monetary gain from selling stolen business information v_b, and (iii) the implementation costs c_i that he incurs when carrying out the attack. To model Cybegangsta's monetary gain from selling customer and employee data v_r, we start with the value of each customer and employee data record, which is $825, as previously defined. To capture the associated uncertainty, we model this with a Uniform $\mathcal{U}(0.8 \times 825, 1.2 \times 825)$ distribution.

Similarly, we can model Cybegangsta's monetary gain from selling business information v_b by starting with the value of each business record, which is $3,000, as previously defined. We capture the associated uncertainty by modelling this with a Uniform $\mathcal{U}(0.8 \times 3,000, 1.2 \times 3,000)$ distribution. Further details on how these models were developed are available in Couce-Vieira et al. (2020b).

As to the implementation costs c_i, we estimate that implementing a social engineering attack would cost Cybegangsta approximately $800 in total.

The impact on Median of a social engineering attack resulting in the loss or exposure of both customer and employee data and business information can be modelled using a similar approach to what we used for employee error. We can represent this as

$$nii_{recexp|a_{soceng}} = 0.1 \times y_{recexp} \times a_{soceng},$$

where a_{soceng} is the number of social engineering attacks that Median is likely to experience in a given year and y_{recexp} follows a Uniform $\mathcal{U}(0, ft_{records})$ distribution, if $a_{soceng} > 0$. We use a lower coefficient of 0.1 in this instance to account for the default portfolio having already been implemented as part of the Small Nation Cyber Essentials. Again, in the case of customer and employee data the number of records $ft_{records}$ is 200,000 and in the case of business information the number of records $ft_{records}$ is 66,000.

Detection costs and probability

We now examine Cybegangsta's expected costs if detected c_d and his probability of detection $hd_{probability}$ (i.e. of being identified as having carried out the attack). As with Antonymous, we can also model Cybegangsta's expected costs if detected c_d through

$$c_d = hd_{probability} \times hd_{cost}.$$

Since we assume that Cybegangsta and Antonymous have relatively equivalent skill levels, we use the same approach as for Antonymous and model $hd_{probability}$ according to a Binomial $\mathcal{B}(a, 0.002)$ distribution. Since a social engineering attack is a one time occurrence, the number of attack days $a = 1$, giving us a Binomial $\mathcal{B}(1, 0.002)$ distribution.

However, one key difference is that the costs if detected hd_{costs} are much higher for Cybegansta, and as a result we model hd_{costs} using a Uniform $\mathcal{U}(1{,}000{,}000; 1{,}300{,}000)$ distribution.

Cybegangsta's utility

We can now build a preference model for Cybegangsta (that is, his random utility). Given the objectives that are relevant for him, Cybegansta's earnings c (i.e. the objectives that he has achieved, calculated in monetary terms) depend on his monetary gain from selling stolen customer and employee data (measured by the amount of money that it can earn by selling one unit of customer and employee data v_r multiplied by the number of stolen customer and employee data records n_r); his monetary gain from selling stolen business information (measured by the amount of money that he can earn by selling one unit of business information v_b multiplied by the number of stolen business records n_b); his implementation costs c_i; and his expected costs if detected c_d. We can aggregate these to obtain

$$c = v_r \times n_r + v_b \times n_b - c_i - c_d.$$

Then, we obtain the distributions of a_{CYB}, b_{CYB}, and ρ_{CYB}, as shown in the Appendix, using a lower bound estimate for Cybegangsta's earnings of $c_* = -11{,}000{,}000$ and an upper bound estimate of $c^* = 22{,}000{,}000$ and assessing that $u(0) \sim \mathcal{U}(0.05, 0.1)$.

Attack probabilities

We now seek to better understand the strategic thinking of Cybegangsta, who must decide whether or not to attack Median. As in the previous examples, we carry out a simulation in order to determine Cybegangsta's optimal attack strategy. For each possible cybersecurity portfolio, we simulate the attack process 1,000 times with just Cybegangsta as the Attacker. This enables us to forecast his expected behaviour, based on the security controls that Median has in place. We present the results of the simulation in Figure 5.4, where, for simplicity, we only present the probability of Cybegangsta carrying out a social engineering attack against Median (or deciding not to attack) given that Median has implemented just the default portfolio.

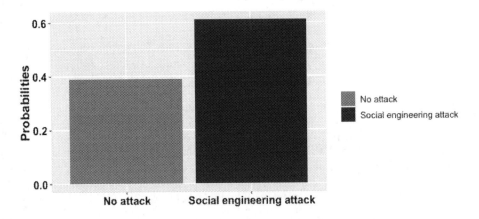

Figure 5.4: Cybegangsta's probability of attacking Median, given that Median has implemented the default portfolio

Modern Republic

Modern Republic is a sophisticated attacker capable of and interested in carrying out both DDoS and social engineering attacks, as reflected in Table 5.7. His objectives include (i) maximising the downtime for Median, as a way to harm Small Nation as well as (ii) stealing business information, as part of his economic espionage activities. In contrast, customer and employee data holds less value for him. Modern Republic is also interested in (iii) minimising his costs if he is detected. The consequences of detection are particularly high for a nation state attacker like Modern Republic, as it is likely to bring economic sanctions and in a worst case scenario could even entail military retaliation. Implementation costs are not a major consideration for Modern Republic, since the attacks are carried out by a military unit funded as part of the government budget. This is reflected in Tables 5.11 and 5.12.

DDoS and social engineering attacks

We can model a DDoS attack carried out by Modern Republic using the same approach that we used for Compeet. The only difference is that Modern Republic is able to wage attacks that are much more powerful, with a capacity of 20 gbps.

As to a social engineering attack perpetrated by Modern Republic, we can model this using the same approach as for Cybegangsta. However, Modern Republic is much more skilled than Cybegangsta, and therefore his probability of successfully tricking Median's employees into falling for a social engineering attack is higher.

The implementation of various security controls—a firewall and internet gateway, an access control system, and/or an intrusion detection system—can reduce the likelihood of a company's employees being fooled by a social engineering attack, as shown in Table 5.9. More specifically:

- The probability of Median's employees falling for a social engineering attack is expected to be 15%, or 0.15. This takes into consideration that the default portfolio consisting of a firewall and internet gateway and an access control system has already been implemented as part of the Small Nation Cyber Essentials.

- If an intrusion detection system is implemented, the probability of Median's employees falling for a social engineering attack is estimated to be 5%, or 0.05. These estimates are based on expert judgement.

Impacts and outcomes

The impacts and outcomes for Modern Republic of perpetrating either a DDoS or a social engineering attack are (i) the value that he derives from the downtime he causes Median v_d, which in turn causes harm to Small Nation and (ii) the value that he derives from the theft of business information v_b, which Modern Republic makes use of for economic espionage. The downtime that Median experiences due to a DDoS attack from Modern Republic can be modelled using the same approach that we used for Compeet (when modelling the duration l of a successful DDoS attack, since these are equivalent), as discussed earlier.

To determine Modern Republic's value per unit of downtime v_d, we must consider that Modern Republic has a tradeoff to consider when it comes to the length of the downtime that he causes Median, as we saw for Antonymous, since increasing this also increases his probability of being detected and his expected costs if detected c_d. To capture the uncertainties surrounding this, we can model Modern Republic's value per unit of downtime v_d using a Uniform distribution, as we did for Antonymous. The value that Modern Republic derives from the theft of business information v_b from Median by carrying out a social engineering attack can be modelled using the same method we used for Cybegangsta.

Detection costs and probability

We now examine Modern Republic's expected costs if detected c_d and his probability of detection, i.e. of being identified as the culprit behind either a DDoS attack or a social engineering attack. To model Modern Republic's expected costs if detected c_d, we must take into account the tradeoff described above where increasing the length of the downtime that Modern Republic causes Median also increases his probability of being detected and his expected costs if detected c_d, as seen with Antonymous as well. We can therefore use a similar approach to that used for Antonymous in modelling Modern Republic's expected costs if detected c_d, enabling us to take its complex relationship with the value per unit of downtime v_d into account.

Given that Modern Republic is a highly skilled attacker, his probability of detection is much lower than that of the attackers in the previous examples. We use expert judgement to estimate this probability at 0.001 for a DDoS attack. This is modelled using a Binomial $\mathcal{B}(a, 0.001)$ distribution, where a is the number of attacks attempted. For a social engineering attack we estimate this probability at 0.0005.

Modern Republic's utility

We can now build a preference model for Modern Republic (that is, his random utility). Given the objectives that are relevant to Modern Republic, Modern Republic's earnings c (i.e. the objectives that he has achieved, calculated in monetary terms) depend on the downtime that he has caused Median (measured in terms of the value expressed in monetary terms that he derives per unit of downtime he causes to Median v_d multiplied by the number of units of downtime that Median experiences d), his value derived from business information (measured in terms of the value expressed in monetary terms that he can obtain from one unit of business information v_b multiplied by the number of stolen business records n_b), and his expected costs if detected c_d. We can aggregate these to obtain

$$c = v_d \times d + v_b \times n_b - c_d.$$

We deduce the distributions of a_{MR}, b_{MR}, and ρ_{MR}, as described in the Appendix, based on a lower bound estimate for Modern Republic's earnings of $c_* = -5{,}000{,}000$ and an upper bound estimate of $c^* = 10{,}000{,}000$ and the assessment that $u(0) \sim \mathcal{U}(0.3, 0.4)$.

Attack probabilities

We now seek to better understand the strategic thinking of Modern Republic, who must choose between four possible attack options concerning Median: (1) carrying out a DDoS attack, (2) carrying out a social engineering attack, (3) carrying out both a DDoS and a social engineering attack, or (4) not carrying out any attacks. As in the previous examples, we carry out a simulation in order to forecast Modern Republic's optimal attack strategy. For each possible cybersecurity portfolio, we simulate the attack process 1,000 times with just Modern Republic as the Attacker. This enables us to forecast Modern Republic's expected behaviour, depending on Median's choices regarding which security controls to implement. We present the results of the simulation in Figure 5.5, revealing the probabilities of all four possible attack options in two different cases: one in which Median has implemented both a DDoS mitigation system and the default portfolio and one in which it has not implemented a DDoS mitigation system but has implemented the default portfolio.

We can see that the implementation of a DDoS mitigation system has some effect in deterring an attack from Modern Republic. However, this is unlikely to be sufficient given Modern Republic's high level of technical skill; Modern Republic may well attack regardless.

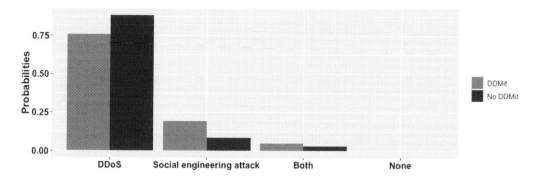

Figure 5.5: Modern Republic's probabilities of attacking Median with various attack types, depending on whether Median has a DDoS mitigation system and assuming Median has implemented the default portfolio

Multiple Attackers

The previous sections examined the probabilities of each individual Attacker carrying out various types of attacks against Median in a given year. However, Median could also be attacked by several of these Attackers during a one year period. To calculate the probability of this, as in Section 4.3.4, we assume conditional independence given the security controls that Median has implemented and estimate

$$p(att_{COMP}, att_{ANT}, att_{CYB}, att_{MR}|port) =$$

$$p(att_{COMP}|port)p(att_{ANT}|port)p(att_{CYB}|port)p(att_{MR}|port),$$

where *port* designates the security controls that Median has implemented and att_x designates the type of attack by Attacker x.

5.4.3   Threat impacts

We now summarise the potential impacts of each threat on Median, evaluating their costs. We also consider other relevant costs as well as gains that Median may have. Then, we use this to build Median's preference model (that is, her utility).

Aggregation of threat impacts

IT systems and computer equipment

Median's IT systems and computer equipment may be impacted as a result of either a fire, a flood, a misconfiguration, or a virus. We present the replacement or repair costs in each instance in Table 5.14.

If Median purchases the conventional insurance product, the insurance will cover 70% of these costs.

Market share

Median's market share is impacted by a DDoS attack from Compeet. The loss in market share that Median experiences is shown in Table 5.15. (As a reminder, the total value of the market is $15,000,000, and Median initially controls 50% of it.)

Threat	Description of impact	Cost
Fire	Replacement	$400 per computer, $1,200 per server
Flood	Replacement	$400 per computer, $1,200 per server
Flood	Repair	$200 per device
Misconfiguration	Replacement	$1,200 per server
Virus	Repair	$31 per device

Table 5.14: Costs of replacement or repair of IT systems and computer equipment for Median

Threat	Description of impact
DDoS	$m \sim 15{,}000{,}000$ min $\lfloor 0.5, l \times R \rfloor$

Table 5.15: Cost of loss in market share for Median

If Median purchases either of the cyber insurance products, the insurance covers 50% of these costs.

Availability

Median can experience a loss of availability (i.e. downtime) due to ransomware, a virus, or a DDoS attack. We present the costs incurred by Median in each instance in Table 5.16.

Threat	Cost	Description of impact
Ransomware	$60,000 per hour	$\Gamma(10, 1)$ per successful attack
Virus	$60,000 per hour	$\Gamma(10, 1)$ per successful attack
Virus	$20 per hour	$w \sim \mathcal{U}(0, 0.05)$, $n_e \times 28 \times w$
DDoS	$60,000 per hour	$l = \sum_j^a l_j$, $l_j \sim \Gamma_{\text{length}}$ if $\Gamma_{\text{gbps}} - 10 > 0$, $l_j = 0$ otherwise

Table 5.16: Costs of loss of availability for Median

The Cyber2 insurance product covers 50% of the impact of this loss of availability due to ransomware, a virus, or a DDoS attack that leaves Median unable to deliver services, as shown in Table 5.10. However, it does not cover the productivity loss caused by a virus.

Customer and employee data, including PII

Another impact on Median is the exposure or loss of customer and employee data, including PII, due to either an employee error, a misconfiguration, a ransomware attack, a virus, or a social engineering attack. We present the costs incurred by Median in each instance in Table 5.17. (As a reminder, the cost of each record is estimated at $825. The number of records exposed or lost depends on the total number of records $ft_{records}$ which is 200,000 in this case.)

Threat	Description of impact
Employee error	$0.1243 \times y_{recexp} \times a_{emperror}$, $y_{recexp} \sim \mathcal{U}(0, ft_{records})$, if $a_{emperror} > 0$
Misconfiguration	$0.1243 \times y_{recexp} \times a_{misconf}$, $y_{recexp} \sim \mathcal{U}(0, ft_{records})$, if $a_{misconf} > 0$
Ransomware	$0.4 \times y_{recexp} \times a_{rsmwr}$, $y_{recexp} \sim \mathcal{U}(0, ft_{records})$, if $a_{rsmwr} > 0$
Virus	$0.1243 \times y_{recexp} \times a_{virus}$, $y_{recexp} \sim \mathcal{U}(0, ft_{records})$, if $a_{virus} > 0$
Social engineering	$0.1 \times y_{recexp} \times a_{soceng}$, $y_{recexp} \sim \mathcal{U}(0, ft_{records})$, if $a_{soceng} > 0$

Table 5.17: Costs of the exposure or loss of customer and employee data for Median

If Median purchases either of the cyber insurance products, the insurance will cover 30% of the cost of the exposure or loss of these customer and employee data records.

Business information

Like for customer and employee data, Median can experience the exposure or loss of business information due to either an employee error, a misconfiguration, a ransomware attack, a virus, or a social engineering attack. We present the costs incurred by Median in each instance in Table 5.18. (As a reminder, the cost of each record is estimated at $3,000. The number of records exposed or lost depends on the total number of records $ft_{records}$ which is 66,000 in this case.)

Threat	Description of impact
Employee error	$0.1243 \times y_{recexp} \times a_{emperror}$, $y_{recexp} \sim \mathcal{U}(0, ft_{records})$, if $a_{emperror} > 0$
Misconfiguration	$0.1243 \times y_{recexp} \times a_{misconf}$, $y_{recexp} \sim \mathcal{U}(0, ft_{records})$, if $a_{misconf} > 0$
Virus	$0.1243 \times y_{recexp} \times a_{virus}$, $y_{recexp} \sim \mathcal{U}(0, ft_{records})$, if $a_{virus} > 0$
Ransomware	$0.4 \times y_{recexp} \times a_{rsmwr}$, $y_{recexp} \sim \mathcal{U}(0, ft_{records})$, if $a_{rsmwr} > 0$
Social engineering	$0.1 \times y_{recexp} \times a_{soceng}$, $y_{recexp} \sim \mathcal{U}(0, ft_{records})$, if $a_{soceng} > 0$

Table 5.18: Costs of the exposure or loss of business information for Median

If Median purchases either of the cyber insurance products, the insurance will cover 30% of the cost of the exposure or loss of these business information records.

Fines

We expect that Median will be fined by the regulator if the exposure or loss of customer and employee data, including PII, is due to either an employee error or an employee falling for a social engineering attack. If the number of records exposed is above 11,000, we anticipate that this fine will be in the $60,000 range. This is not covered by any of the insurance options, as it is not legal for insurance to cover fines in Small Nation.

Other costs and gains

We also consider other costs as well as gains that Median may have. These involve the costs of the security controls and of the insurance product(s) that Median decides to purchase. This also consists of any compensation that Median may obtain if attacked, in the form of an insurance payout.

Median's preference model

We now build Median's preference model (which is determined by her utility). In order to be able to determine Median's utility, we have translated all of the impacts on Median described above into monetary terms, if not already expressed that way. We define:

- the *costs of equipment* as the total replacement and repair costs of IT systems and computer equipment (denoted as q),

- the *loss in market share* as the monetary value of the decrease in market share that Median experiences (denoted as m),

- the *loss from customer and employee data, including PII* as the monetary cost to Median of the exposure or loss of one unit of customer and employee data, including PII (denoted as w_r),

- the *loss from business information* as the monetary cost to Median of the exposure or loss of one unit of business information (denoted as w_b),

- the *loss due to downtime* as the monetary loss for Median per unit of downtime that it experiences (denoted as w_d),

- *fines* as the cost of the fines levied on Median (denoted as f),

- the *cost of security controls* as the cost of the security controls that Median decides to purchase (denoted as c_s),

- the *cost of insurance* as the cost of the insurance products that Median decides to purchase (denoted as c_i), and

- the *compensation from insurance*, or the amount of money that Median gains from an insurance payout (denoted as g_i).

These are summarised in Table 5.19.

Impact in monetary terms	Variable
Damage to equip	q
Loss in market share	m
Customer & employee data loss	w_r
Business info loss	w_b
Downtime	w_d
Fines	f
Costs of sec controls	c_s
Cost of insurance	c_i
Compen from insurance	g_i

Table 5.19: Impacts on Median in monetary terms

We also need to define variables to capture the total number of units. We therefore create variables to express the *number of customer and employee records, including PII* (denoted as n_r), the *number of business records* (denoted as n_b), and the *number of downtime units* (denoted as d). These are summarised in Table 5.20.

Impact in total units	Variable
Customer & employee data units	n_r
Business info units	n_b
Downtime units	d

Table 5.20: Impacts on Median in total units

We then aggregate these threat impacts, now that they have been monetised, using the multiattribute cost function

$$c_D = q + m + (w_r \times n_r) + (w_b \times n_b) + (w_d \times d) + f + c_s + c_i - g_i$$

where c_D are the costs of the attack for the Defender. This is a generalisation of a model presented in Couce-Vieira et al. (2020a).

As a general utility function we adopt

$$u_D(c_D) = a_D(1 - \exp(\rho_D c_D)) + b_D$$

with a_D, b_D, and $\rho_D > 0$. The best case scenario for Median is to have no costs, or $c_D^* = 0$ and we assign to it $u_D(c_D^*) = 1$. The worst case scenario for Median is estimated at $c_{D*} = 8{,}000{,}000$ and we assign to it $u_D(c_{D*}) = 0$. We also assess $u_D(4{,}000{,}000) = 0.7$ using the probability equivalent method (Farquhar, 1984). Simple computations similar to those in the Appendix lead to the estimates $a_D = 0.2083$, $b_D = 1$, and $\rho_D = 0.2083$. Therefore the utility function that we can adopt for Median is

$$u_D(c_D) = 0.2083(1 - \exp(0.2197 \times c_D)) + 1.$$

5.5 Model solution and assessment

5.5.1 Number of possible cybersecurity portfolios

Now that we have assessed all of the components of the model, we seek to determine Median's optimal cybersecurity portfolio, which is made up of a combination of security control portfolios and insurance portfolios. Median effectively only has seven security controls to decide on, given that of the nine security controls in Table 5.9, it has already incorporated two of them in its default portfolio to comply with the Small Nation Cyber Essentials. This, in turn, means that there are 128 possible combinations of security control portfolios that Median could select.

Median has three insurance products to choose from, but there are only six possible combinations of insurance portfolios that Median might select, as it would not make sense for Median to pick insurance portfolios that contain both cyber insurance products. This means that there are 768 possible cybersecurity portfolios for Median in total (calculated by multiplying 128×6). However, only 719 of these are within Median's budget of \$60,400 that remains after the default portfolio has been implemented. Table 5.21 shows the number of potential security control portfolios that are within Median's budget for each of its six possible insurance portfolios.

Insurance portfolio	Security control budget	Number of security control portfolios
None	60,400	128
Conv	57,400	128
Cyber1	53,400	126
Cyber2	48,400	113
Conv+Cyber1	50,400	120
Conv+Cyber2	45,400	104

Table 5.21: Number of potential security control portfolios for each insurance portfolio

5.5.2   Optimisation results

As it is computationally manageable, we run a simulation for each of these 719 cybersecurity portfolios to determine their expected utilities, as described in Section 4.3.4. We have chosen a sample size of 1,000. First, we present the results when we consider a potential attack from just one of the targeted Attackers (taking all six non-targeted threats into account as well). We do this for each Attacker, showing the findings in Tables 5.22 to 5.25. We then present the results when we consider all four targeted Attackers (and the other six non-targeted threats), displayed in Table 5.26. In each table, we present the five best cybersecurity portfolios for Median for the given scenario. In each case, we show the security control portfolio (with a "1" if the security control is included and a "0" otherwise); the insurance portfolio; the total cost of the investment (i.e. the combined cost of the security controls and the insurance, not including the cost of the default portfolio), measured in US dollars; the expected annual cybersecurity operational costs, measured in millions of US dollars; and Median's expected utility.

Compeet and Antonymous

We present the case in which Compeet is the only targeted Attacker in Table 5.22 and the case in which Antonymous is the only targeted Attacker in Table 5.23 (taking the other six non-targeted threats into account as well).

Sprk	FD	DDMit	SecCnf	MPS	PVM	IDS	Insurance	Investment	Operational costs	Utility
0	0	1	1	0	0	0	Conv	16,000	0.0167	0.9992
0	0	1	0	0	1	0	Conv	16,600	0.0204	0.9990
0	1	1	1	0	0	0	None	17,800	0.0241	0.9988
0	1	1	1	0	1	0	Conv	22,400	0.0245	0.9988
0	0	1	1	1	1	0	Conv	21,600	0.0248	0.9988

Table 5.22: Median's best portfolios when Compeet is the only Attacker

Sprk	FD	DDMit	SecCnf	MPS	PVM	IDS	Insurance	Investment	Operational costs	Utility
0	0	1	1	0	1	0	Conv	17,600	0.0179	0.9991
0	0	1	0	1	0	0	Conv	19,000	0.0198	0.9990
1	0	1	1	1	0	0	Conv	20,600	0.0211	0.9990
1	0	1	0	1	1	0	Conv	21,200	0.0218	0.9989
0	1	1	0	0	1	0	Conv	21,400	0.0236	0.9988

Table 5.23: Median's best portfolios when Antonymous is the only Attacker

Compeet and Antonymous have relatively similar expected behaviours, as we observed in Figures 5.2 and 5.3, and their attacks are less harmful than those of Cybegangsta or Modern Republic. To defend against an attack by either of them, Median should clearly invest in a DDoS mitigation system, as it features in the five best portfolios in the case of either Compeet or Antonymous being the only Attacker. It should also purchase the conventional insurance product, as this appears in the vast majority of the optimal cybersecurity portfolios. Investing in a secure configuration would be particularly beneficial if Compeet is the only Attacker, as it is present in four out of five of the optimal cybersecurity portfolios. (This is in addition to the default portfolio that has already been implemented consisting of a firewall and internet gateway and an access control system.)

Cybegangsta and Modern Republic

We present the case in which Cybegangsta is the only targeted Attacker in Table 5.24 and the case in which Modern Republic is the only targeted Attacker in Table 5.25 (taking the other six non-targeted threats into account as well).

Sprk	FD	DDMit	SecCnf	MPS	PVM	IDS	Insurance	Investment	Operational costs	Utility
0	1	0	0	1	0	1	Cyber1	45.800	2.3183	0.6346
0	0	1	1	0	1	1	Conv+Cyber2	59.600	2.5077	0.6304
1	0	1	1	0	0	1	Conv+Cyber1	53.600	2.3777	0.6228
1	1	0	1	0	0	1	Conv+Cyber2	51.400	2.3849	0.6213
0	1	0	1	1	1	1	Conv+Cyber2	56.400	2.3875	0.6162

Table 5.24: Median's best portfolios when Cybegansta is the only Attacker (includes non-targeted threats as well)

Sprk	FD	DDMit	SecCnf	MPS	PVM	IDS	Insurance	Investment	Operational costs	Utility
1	1	1	0	0	0	1	Cyber2	59.400	0.6158	0.9449
1	0	1	1	0	0	1	Cyber2	55.600	0.6346	0.9429
0	1	1	0	0	0	1	Cyber2	58.800	0.6190	0.9418
1	0	1	1	0	0	1	Conv+Cyber2	58.600	0.6514	0.9385
0	0	1	0	1	1	1	Cyber2	59.600	0.6915	0.9357

Table 5.25: Median's best portfolios when Modern Republic is the only Attacker (includes non-targeted threats as well)

Compared to Compeet or Antonymous, an attack by either Cybegangsta or Modern Republic could have a much more damaging effect on Median. To adequately defend against an attack by either of them, Median needs to invest significantly more financial resources. Median should clearly purchase an intrusion detection system, as this features in the five best cybersecurity portfolios when either Cybegangsta or Modern Republic is the only Attacker. In addition, investing in a secure configuration would be especially useful if Cybegangsta is the only Attacker, being present in four of the top five optimal cybersecurity portfolios. A DDoS mitigation system would be highly beneficial if Modern Republic is the only Attacker, as this appears in the five best portfolios. (This is in addition to the default portfolio that has already been implemented consisting of a firewall and internet gateway and an access control system.) Moreover, the insurance portfolios that are needed when either Cybegangsta or Modern Republic is the only Attacker predominantly contain cyber insurance products. In contrast, those needed when Compeet or Antonymous is the only Attacker require conventional insurance products at most.

All four potential Attackers

We now present the full analysis in which all four targeted Attackers are present (and the other six non-targeted threats as well) in Table 5.26. These findings have been corroborated by a sensitivity analysis.

Sprk	FD	DDMit	SecCnf	MPS	PVM	IDS	Insurance	Investment	Operational costs	Utility
1	0	1	0	0	1	1	Cyber2	56.200	3.1034	0.4698
1	0	1	1	0	1	1	Conv+Cyber2	60.200	3.0002	0.4693
0	0	1	0	1	1	1	Cyber2	59.600	2.9579	0.4678
1	0	1	0	0	1	1	Conv+Cyber2	59.200	3.0308	0.4674
1	0	1	1	0	0	1	Cyber2	55.600	3.0690	0.4603

Table 5.26: Median's best portfolios when all threats are considered (targeted Attackers and non-targeted threats)

We can see that defending against all four possible Attackers requires a significant investment. Given that the optimal cybersecurity portfolio is the one with the maximum expected utility, in this case such portfolio contains the following controls: a DDoS mitigation system, an intrusion detection system, a sprinkler system, and a patch vulnerability management system (as well as the default portfolio, which consists of a firewall and internet gateway and an access control system). It also contains the Cyber2 insurance product.

It is useful to also consider the next four cybersecurity portfolios, particularly since their expected utilities are very close to those of the optimal portfolio.[4] They predominantly include the same security controls as the optimal portfolio. However, we also observe that by adding a secure configuration and a conventional insurance product, we can obtain a more comprehensive solution for just a slightly higher cost. This could therefore be very beneficial for Median.

5.6 Conclusions

This chapter has shown how to implement the model we developed in Chapter 4 to support companies in their CSRM decisions, helping them determine their optimal cybersecurity portfolio, or the best way for them to allocate their budget between spending on security controls and on cyber insurance products. The case study presented here serves as a blueprint for companies who wish to use the model, illustrating how to formulate it, how to assess the various components, and how to solve for the optimal cybersecurity portfolio. The model can be modified or expanded, based on the characteristics of the company it is being applied to.

---

[4]The standard deviations are approximately 0.008, based on the sample sizes of 1,000.

The case study also helps illustrate how to implement the other models we developed in Chapter 4 to assist organisations and insurers with their decisions involving risk management in cybersecurity. The same types of methods shown here to formulate, assess the components, and solve the model can be applied to these models as well.

Appendix

This section illustrates how we model the Defender's beliefs about the Attacker's parameters using the distributions of a_A, b_A, and ρ_A. Assuming that an Attacker is constant risk prone, we represent his utility function as

$$u_A(c) = a_A \times \exp(\rho_A \times c) + b_A.$$

We estimate the Attacker's lower bound $c_* < 0$ and upper bound $c^* > 0$ values and associate them with utilities of 0 and 1, respectively. We then assess the Attacker's utility u for $c = 0$, leading to

$$a_A \times \exp(\rho_A \times c_*) + b_A = 0,$$
$$a_A \times \exp(\rho_A \times c^*) + b_A = 1,$$
$$a_A + b_A = u,$$

or, equivalently,

$$b_A = u - a_A, \tag{5.1}$$

$$\rho_A = \frac{1}{c_*} \log \frac{-b_A}{a_A}, \tag{5.2}$$

$$a_A \left(\frac{u - a_A}{a_A} \right)^{\frac{c^*}{c_*}} + (u - a_A) = 1.$$

If we choose c_* and c^* such that $\frac{c^*}{c_*} = -2$, then

$$(3u - 1)a_A^2 + (2u - 3u^2)a_A + (u^3 - u) = 0.$$

We then deduce a_A through

$$a_A = \frac{(3u^2 - 2u) \pm \sqrt{(2u - 3u^2)^2 - 4(3u - 1)(u^3 - u)}}{2(3u - 1)}. \tag{5.3}$$

This, in turn, enables us to calculate b_A and ρ_A.

However, since we do not have sufficient information about the Attacker to properly elicit u, we typically assess it using a Uniform $\mathcal{U}(u_1, u_2)$ distribution. We then deduce the corresponding distributions of a_A, b_A, and ρ_A, which form the basis of the random utility model for the Attacker.

As an example, assume we use $c_* = -2.5$ million, $c^* = 5$ million, and the uncertainty surrounding the Attacker utility u is modelled using a Uniform $\mathcal{U}(0.1, 0.3)$ distribution. Then, we sample 1,000 observations from U and, using (5.3), (5.1), and (5.2), we obtain the distributions in Figure 5.6 showing the histograms of the random parameters A_A, B_A, and P_A.

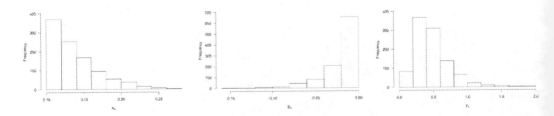

Figure 5.6: Histograms of A_A, B_A, and P_A

Then, the Attacker's random utility from the Defender's perspective would be

$$U_A(c) = A_A \times \exp(P_A \times c) + B_A.$$

This random utility function can then be used to simulate from the Attacker's strategic thinking in order to forecast his actions, as explained in the main text.

6

Conclusions

Caroline Baylon
AXA

Deepak Subramanian
AXA

Jose Vila
DevStat and University of Valencia

David Rios Insua
ICMAT

CONTENTS

This concluding chapter opens with a review of the key concepts presented in this book, including a discussion of the decision-making and cybersecurity risk management challenges that companies and insurers face, why current approaches are not effective, and how the models we have developed can help overcome these issues. We then highlight some of the new and innovative contributions that this book makes to the cybersecurity and cyber insurance fields. This includes research into areas that have up until now received insufficient attention, notably, decision-making challenges for Small and Medium Enterprises. In addition, we explored how to leverage behavioural economics insights to nudge better cybersecurity behaviour. We also developed models that make use of Adversarial Risk Analysis and that enable dynamic cyber insurance pricing and provide a better understanding of accumulation risk. Finally, we discuss the policy implications of our work.

6.1   Introduction

This book has described some of the challenges that organisations face when it comes to decision-making involving cybersecurity and cyber insurance and presented novel solutions to these decision-making challenges in the form of new mathematical models. It achieves this by bringing together multidisciplinary perspectives, drawing on a wide range of fields from psychology and behavioural economics to decision analysis, statistical risk analysis, and game theory. We have employed a number of diverse methodological approaches including modelling techniques, interviews, and experiments. These were developed and carried out as part of a two year research project funded by the European Union called the CYBECO (Supporting Cyber Insurance from a Behavioural Choice Perspective) Project, of which this book is the culmination.

In this concluding chapter, we first review the key concepts presented in this book. We then discuss some of the novel and innovative contributions that this book makes to the cybersecurity and cyber insurance fields, including key areas that have been underinvestigated thus far in the literature. Lastly, we review the policy implications of our work.

6.2   The key concepts

This section summarises the key concepts in this book. First, we detail the decision-making challenges that companies and insurers must contend with when it comes to cybersecurity. Next, we discuss why current approaches for Cybersecurity Risk Management (CSRM) are not effective, including why cyber insurance is not yet widely used as a viable CSRM solution. We then briefly review the models we have developed in this book and discuss how they help overcome these issues. We also describe the tools we have created to make it easier for an organisation to implement these models.

6.2.1   Cybersecurity decision-making challenges for companies and insurers

In the first three chapters of this book, we described some of the decision-making challenges that organisations face when it comes to cybersecurity and cyber insurance. We first examined the decision-making process at the organisational level (Chapter 2), then turned to individual decision-making within organisations (Chapter 3). The challenges for companies include budget constraints, requiring them to make trade-offs when it comes to spending on cybersecurity and cyber insurance. Many also have an inadequate understanding of the cybersecurity risks they are facing. The problem is exacerbated by the rapidly evolving cyber threat landscape, with the number of attacks and the sophistication of attackers rising precipitously. Company decision makers also lack knowledge about available cyber insurance products. Many report confusion about what cyber insurance policies will cover in the event they experience a cyber attack. These factors all make it difficult for companies to decide upon the cybersecurity and cyber insurance products that they need to buy, which often results in underinvestment in these areas.

We also described some of the decision-making challenges that insurance companies must contend with (Chapter 1). Insurers struggle to accurately assess cybersecurity risk, making it difficult for them to develop (and price) cyber insurance products. This is partly due to the dynamic nature of the cyber threat landscape mentioned above. Another factor is the limited amount of cyber incident data available, stemming from organisations' reluctance to disclose that they have been attacked. This is a major challenge for insurers given that they have traditionally relied upon historical data for underwriting. A related issue is accumulation risk, or the risk that a single event could spread to multiple lines of business, resulting in unexpectedly large claims for insurers. Another problem is moral hazard, or the risk that an insured company will engage in riskier behaviour knowing that the insurer will cover the costs. It is also difficult for underwriters to estimate intangibles such as the loss of brand value. These challenges are compounded by an acute shortage of experienced cybersecurity underwriters.

6.2.2 Issues with current cybersecurity risk management approaches

For companies, the use of reliable CSRM methods is essential in order to manage these threats. However, current CSRM methods have a number of drawbacks, many of them concerning the way threats are modelled, as discussed in detail in Chapter 1. These methods often rely on risk matrices, which assign the same ratings to threats that are significantly different. This may be a particular concern given the increasing variety of threats. They also do not typically take into account the intentionality of threats. These drawbacks are likely to result in a sub-optimal allocation of all too frequently scarce cybersecurity resources.

Given these issues with current CSRM methods, cyber insurance can form an important part of an alternative CSRM strategy for companies. For one, it enables them to partially transfer the cybersecurity risk to insurance providers. Cyber insurance also provides companies with cyber threat information and assistance with cyber incident response. Yet despite its benefits, cyber insurance is still limited and uptake has not reached expectations, in large part due to the decision-making challenges for both companies and insurers described in the previous section.

This is all the more regrettable in that, beyond providing companies with a viable CSRM solution, cyber insurance also has the potential to reduce overall cybersecurity risk for society as a whole. Cyber insurance can incentivise companies to improve their cybersecurity levels, either in exchange for reduced premiums or by requiring them to implement certain cybersecurity measures as a condition of coverage. Greater uptake of cyber insurance by companies also makes it possible for insurers to acquire more cyber incident data, leading to improvements in the ability to assess cyber risk in a "virtuous cycle".

6.2.3 A solution in the form of new mathematical models

To address the problems faced by both companies and insurers, we developed a series of new models to deal with risk management problems in cybersecurity, which we presented in Chapter 4. These models make use of Adversarial Risk Analysis (ARA), a novel subfield of decision analysis that draws on statistical risk analysis and game theory and takes the actions of adversarial threat actors into account, making use of Multi-Agent Influence Diagrams (MAIDs) as part of the process.

The key model is a model to assist companies in deciding upon their cybersecurity resource allocation, which we refer to as "the CSRM model". By using ARA to more accurately model threats, this new model offers considerable improvements over current risk matrix-based CSRM methods. It also utilises cyber insurance as a major part of the CSRM process, as the model also advises companies on their selection of a cyber insurance product. More specifically, the model helps companies determine their optimal budget allocation strategy between investing in security controls (to defend against attacks) and in cyber insurance products (to receive a payout in the event of an attack). We refer to this combination of security controls and cyber insurance as the organisation's "cybersecurity portfolio".

We also presented a series of auxiliary models to assist insurers. Some of these models aim to help insurers determine the optimal design of cyber insurance products. This includes a model to optimise price and coverage (the maximum price that a client would be willing to pay and the minimum coverage that they would accept). We also describe how to carry out market segmentation, or determining clusters of organisations that would choose similar cybersecurity portfolios, to facilitate marketing of the products. In addition, we discuss how to engage in dynamic pricing of cyber insurance products, i.e. for premiums to automatically adjust in response to an increase or decrease in an insured company's cybersecurity risk.

The series of models for insurers also includes models to help them deal with fraud. This includes a model for helping an insurer decide whether to issue a cyber insurance policy to a customer, given the possibility that the customer might commit fraud. It also includes a model for helping the insurer determine whether it should classify a particular claim by a customer as fraudulent. Finally, we presented a model to assist insurers in determining the level of reinsurance that they need to take out. This model helps to better understand accumulation risk.

6.2.4   Facilitating the implementation of these models

To simplify the process for companies wishing to use the CSRM model, as part of the CYBECO Project we developed a prototype "toolbox" that provides an online interface, which is described in Chapter 3. In response to information provided by the company, the toolbox outputs an assessment of its cybersecurity risk as well as advice on the optimal allocation of financial resources based on the CSRM model.

For companies that want to carry out a more detailed assessment, we also presented a full case study in Chapter 5 showing how to implement the CSRM model. This involved illustrating how to model each of its various components and how these can be adapted for the specific characteristics of a particular company. The same approach can also be applied to implement the auxiliary models we have developed to assist insurers. In addition, we have made the R code used in the case study available for others to make use of.

6.3   Novel and innovative contributions to cybersecurity and cyber insurance

In this section we highlight some of the new and innovative approaches to cybersecurity and cyber insurance that are presented in this book. One of the book's major contributions is its behavioural approach, seeking to better understand the behavioural choices of organisations (Defenders), threat actors (Attackers), and insurers. As part of this, we explore an area which has not received sufficient research attention: organisational decision-making when it comes to cybersecurity and cyber insurance, with a focus on Small and Medium Enterprises (SMEs). We also draw on behavioural economics to obtain insights regarding how to "nudge" organisations to engage in better cybersecurity behaviour. Most importantly, we make use of Adversarial Risk Analysis in order to model the behaviour of various threat actors as well as to determine how organisations should behave given their likelihood of being attacked. In addition, we apply ARA to help inform the behavioural choices of insurers. The development of models for dynamic pricing of cyber insurance and for better understanding accumulation risk are another important contribution of this book.

6.3.1   A behavioural approach

A key contribution of this volume has been to examine the behavioural choices of key players: organisations (Defenders), threat actors (Attackers), and insurers. In particular, we have gained a greater understanding of their decision processes and how these relate to cybersecurity risk. This, in turn, has enabled us to develop innovative methods of risk management in cybersecurity.

Shedding light on organisational decision-making for cybersecurity and cyber insurance

When analysing the behavioural choices of decision makers within organisations, we looked at the factors behind their risk reduction and risk transfer decisions as regards to investing in security controls and taking out cyber insurance. Up until now limited research has been conducted on how companies make decisions involving cybersecurity and cyber insurance (with organisational, behavioural, and environmental perspectives especially overlooked). Yet these decisions are extremely important for companies, not least because they can involve a sizeable part of companies' increasingly squeezed budgets.

To address this gap, we conducted a study in which we interviewed practitioners inside companies, including those responsible for cybersecurity decisions, to identify the factors behind cybersecurity and cyber insurance decision-making, drawing on the Burke and Litwin Performance and Change Model (Chapter 2). The study found considerable variation between companies in terms of how decisions are taken, with numerous (non-universal) boards, committees, and departments, each with their own processes. Other findings included the importance of external influences on company decision-making, with legislation playing a key role in driving cybersecurity and cyber insurance uptake, a finding consistent with the Burke and Litwin Model.

Putting a spotlight on SMEs

The cybersecurity challenge may be particularly acute for SMEs, yet the sparse research that has been done on organisational decision-making involving cybersecurity and cyber insurance has tended to ignore SMEs in favour of large companies. This is a concern because SMEs may be particularly vulnerable to cyber attacks: They have relatively fewer financial resources to invest in cybersecurity. They also generally have less cybersecurity expertise and tend to underestimate the risk that cyber attacks pose to them, with many thinking that they have a low probability of being attacked. Yet in actuality, SMEs are a major attack target for cybercriminals.

To bridge this gap, we carried out a study to examine how SMEs decide on cyber insurance adoption, including conducting interviews with SME representatives (Chapter 2). Overall, we found that the cyber insurance decision-making process for SMEs is problematic due to a poor understanding of cybersecurity risks and the dynamic nature of those risks. We developed a Cyber Insurance Adoption Model for SMEs to illustrate the SME decision-making process involving cyber insurance, which extends traditional Protection Motivation Theory (PMT) with additional elements.

We also concentrated on SMEs in the Behavioural Economics Experiment (BEE) that we conducted (Chapter 3), which is described more fully in the next section. The experiment studied the cybersecurity decision-making of SME employees.

Finally, we designed some of the models with SMEs in mind. When building the CSRM model, we used an SME as the use case, recognising that these decisions are particularly critical for budget-conscious SMEs (Chapter 4). The toolbox is aimed at SMEs (Chapter 3), and the case study detailing how to implement the CSRM model is also built around an SME (Chapter 5).

Leveraging behavioural economics to better understand the behaviour of organisations

In addition to the organisational decision-making studies described above, we also sought to better understand the behaviour of decision makers within organisations by drawing on behavioural economics, which applies psychological insights to human behaviour and is based on the premise that decision-making is not always rational (Chapter 3). This included making use of Dual-Thinking Theory, which views all decisions as being made based on both fast, subconscious System 1-type thinking and slow, conscious, rational System 2-type thinking, and of Prospect Theory, which assumes that people value losses and gains very differently and that they are particularly risk averse when it comes to losing what they already have.

The use of behavioural economics techniques also gave us insights into how to "nudge" better cybersecurity behaviour. Since System 1-type thinking is subliminal, it is not possible to interview decision makers to assess how their cybersecurity decision-making will respond to behavioural nudges appealing to System 1-type thinking. Instead, we conducted a Behavioural Economics Experiment (BEE), which allowed us to observe users' actual cybersecurity behaviour in a controlled environment in response to various nudges. It also enabled us to test and refine the toolbox that we developed based on the CSRM model. The experiment involved SME employees responsible for cybersecurity, as mentioned earlier, who were asked to make decisions regarding the cybersecurity budget allocation of a fictional company, based on a ranking of different options presented using the toolbox's online interface. We identified a number of nudges that had a significant impact on subjects' cybersecurity decision-making. Subjects were more likely to choose the top option when a message indicating it was recommended by cybersecurity experts was displayed, along with a link to make the purchase. Presenting the different options in terms of expected values in the event of a cyber attack also induced the subjects to choose the top option. The effect was especially strong when the expected values were framed in terms of expected losses, which is in line with Prospect Theory. In this way, we were able to assess how various nudges could get users to pay more attention to the outputs of the toolbox and its underlying CSRM model.

Making use of Adversarial Risk Analysis

Using ARA to forecast the behaviour of threat actors and help organisations determine how they should behave

We also examined the risk generation decisions of threat actors, seeking to understand and make forecasts about their behaviour and likelihood of deciding to attack an organisation. This approach has enabled us to overcome problems with current CSRM methodologies that typically do not take into account the intentionality of threats (Chapter 1).

As previously mentioned, the CSRM model and the other models that we have developed to deal with risk management problems in cybersecurity make use of a novel technique, Adversarial Risk Analysis (Chapter 4). The ARA methodology allows us to forecast the behaviour and actions of the different threat actors that may wish to attack a particular organisation (Chapter 4). By considering the threat actors' goals and uncertainties, the use of ARA examines the threat actors' strategic thinking to determine their probability of attacking the organisation. This makes it possible to better assess the risk that the different threat actors pose to the organisation.

The organisation can then determine how it should behave given the threat actors' forecasted behaviour, enabling it to decide on its best strategy given its likelihood of being attacked by different threat actors. This allows it to better prioritise when it comes to guarding against these threats. The organisation can make use of the CSRM model that we have developed based on ARA, helping it determine its optimal cybersecurity resource allocation, or the best way for it to allocate its budget between investing in security controls, to reduce the risk of an attack, and purchasing cyber insurance, to transfer the risk.

Applying ARA to inform the behavioural choices of insurers

We also take into account the behavioural choices of insurers in terms of their risk assessment decisions. Notably, we discussed the factors that insurers consider when underwriting policies, including the information they use to assess a company's risk profile, how they model the risk in the domain as a whole, the factors they take into account when setting premiums, and the instruments they use in designing the policy and drafting the contract (Chapter 4). To help insurers better assess the risk, they can make use of the series of auxiliary models that we have developed based on ARA, which can assist them with issues involving risk management in cybersecurity such as the optimal design of cyber insurance products and determining whether to issue a policy, whether a claim is fraudulent, or the level of reinsurance needed.

6.3.2 Enabling dynamic cyber insurance pricing

As described above, we also proposed a model to price insurance products dynamically, that is, to enable premiums to automatically adjust in response to an increase or decrease in an insured organisation's cybersecurity risk (Chapter 4). This has been at the forefront of insurers' preoccupations for some time, given that an organisation's risk profile at the time it takes out an insurance policy may not be the same several months later. It is thus important to be able to adjust the pricing.

This is increasingly possible to do due to the emergence of companies that gather real time information about an organisation's IT infrastructure, security environment, etc. to obtain a clear picture of its cybersecurity risk at any given point in time. From a modelling perspective, we can then apply a discount if the organisation's cybersecurity risk remains below a certain level for a given period of time, and a penalty if it reaches that level and the organisation does not take action to improve its cybersecurity.

6.3.3 Achieving a better understanding of accumulation risk

As previously mentioned, we have proposed a model to better understand accumulation risk, or the risk that a single event could spread to multiple lines of business (Chapter 4). This can lead to an unexpectedly large number of insurance claims and substantial losses for insurers. Accumulation risk is especially challenging in the cyber realm: In the physical world, an event such as a natural disaster (e.g. a hurricane or an earthquake) may trigger a surge in claims, but these are limited to a particular geographic area. A cyber attack, by contrast, can result in claims anywhere around the world. For example, the WannaCry cyber attack infected some 250,000 users in 150 countries. There is particular concern surrounding companies' growing reliance on cloud and other third party services, as an attack on a cloud service provider could result in widespread business interruption for thousands of companies.

The model does so by detailing different market segments, which then makes it possible to understand the accumulation effect of a cyber attack that affects a given market segment. For example, it is possible to gain greater insight regarding the potential impact of a ransomware attack hitting hospitals and the wider medical sector.

6.4   Policy implications

We now turn to the policy implications of our work. On the cyber insurance side, many of our recommendations focus on measures to foster trust between companies and insurers, which is key for the further development and uptake of cyber insurance. We also propose a number of measures to strengthen cybersecurity, which will help create the conditions necessary for cyber insurance to advance further. Lastly, we look at how to maximise the impact of policy measures to boost cybersecurity and cyber insurance.

6.4.1   Policy measures to foster trust between companies and insurers

Standardising cyber insurance policy wording

Policy measures to foster trust between companies and insurers are key for cyber insurance to be taken up on a large scale. We found that mistrust between companies and insurers is a major obstacle to the widespread adoption of cyber insurance, with companies expressing concern that they may not be properly covered and indemnified (Chapter 2).

In order to help reassure companies, a policy measure to promote the use of standard language for cyber insurance policies would be beneficial. In particular, there is a desire for more detailed cyber insurance policy wording regarding the specific terms and conditions of coverage, e.g. inclusions and exclusions. This would also address companies' confusion over what policies cover and help them better understand what residual risks they can transfer with cyber insurance. However, given the rapidly changing nature of the cybersecurity environment, policy measures to standardise language need to consider the importance of finding the right balance between providing enough detail to reassure companies and giving insurers enough room to take into account new developments in cybersecurity risk. Further research is needed to investigate the most appropriate level of specificity in the wording. This policy measure could be accomplished through legislation or other means.

Reinforcing the role of insurers as cybersecurity advisors

Another policy measure to build companies' trust in insurers involves solidifying the role of insurers (and brokers) in advising their customers on cybersecurity, including by regulating their duties and liability (Chapter 2). Since cyber insurance policies also involve providing information on cybersecurity threats, assistance with cyber incident response, and other cybersecurity advice, this will give companies confidence that they are receiving high quality advice from insurers and brokers.

It will also ensure that insurers and brokers fully understand their responsibility as advisors (Chapter 2). In order to bolster their advice, insurers and brokers should also be encouraged to make use of robust modelling techniques, such as the CSRM model we have developed (Chapter 4). They may also wish to use its accompanying toolbox (Chapter 3). Such policy measures would boost the import and value of insurers' cybersecurity advice, helping to strengthen cybersecurity more broadly within the ecosystem.

Promoting risk selection to reduce the moral hazard problem

Policy measures can help increase insurers' trust in companies as well. This includes policy measures to promote the use of risk selection in cyber insurance, in order to reduce the moral hazard problem (Chapter 2). The study involving the Agent-Based Model found that, when designing policy measures to increase cyber insurance uptake, it is essential to keep the moral hazard problem in check through risk selection. One way this might be accomplished is by promoting the use of models such as those we have developed for dealing with cyber insurance fraud, notably the model for determining whether to issue a cyber insurance policy given the possibility of fraud (Chapter 4). This can play an important role in increasing insurers' trust in prospective customers via the use of risk selection. Policy measures to increase the cybersecurity of organisations more broadly will also help achieve this. These are discussed in the following section.

6.4.2 Policy measures to strengthen cybersecurity

Promoting cybersecurity certification schemes

Given that many organisations fail to accurately assess their cybersecurity risk, policy measures to raise awareness of cybersecurity best practices or establish or promote existing cybersecurity certification schemes can play a key role in helping them better understand their cybersecurity readiness (Chapter 2). In our interviews of SME personnel we found that they are often not willing to share their cybersecurity methods, because they view the cyber threats they face and the mitigation tactics they use as sensitive topics, slowing down the adoption of cybersecurity best practices. There is therefore an important role for cybersecurity certification schemes in increasing knowledge about best practices and encouraging their uptake. In order to encourage organisations to comply with cybersecurity certification schemes, regulations could specify that certification helps them meet their duty of care and means that they are less likely to be fined or held liable in case of a breach.

Enhancing information sharing

The scarcity of good cyber incident data makes policy measures to improve information sharing of the data that is available all the more important (Chapter 2). We have discussed a number of reasons for companies' reluctance to disclose cyber incidents in this book, many stemming from the fear of reputational damage. New EU cybersecurity directives like GDPR are generating a lot of data surrounding data breaches and cyber incidents. Discussions are currently underway about whether the data collected by government agencies under GDPR could be made more widely available. This could help insurers improve their cybersecurity risk assessments as well as aid organisations in better understanding their cybersecurity risk. To do so, stringent measures to ensure the anonymisation of the data would be needed. In the meantime, we can use structured expert judgement methods such as those described in Chapter 5 to help compensate for limited data.

Imposing further financial costs on companies that experience cyber incidents

There is a need for further policy measures and/or regulations that impose financial costs on companies for cyber incidents, such as the cost of notifying affecting organisations or individuals following a breach or fines in the event of a breach attributable to non-compliance with regulations (Chapter 2). This will not only increase companies' cybersecurity readiness but is also likely to encourage cyber insurance uptake in countries where it is legally permissible for insurance to cover fines for cyber incidents. In the US, the 1996 Health Insurance Portability and Accountability Act (HIPAA), which imposed fines on healthcare providers that violate security standards, and the 2003 California Data Breach Laws, which required companies to notify California residents if their personal information is exposed in a data breach, significantly contributed to the growth of the US cyber insurance market. However, there is considerable confusion surrounding the situations in which fines are insurable. There is therefore a need for countries who have not yet done so to clarify their positions in this regard.

6.4.3   Maximising impact

Moving away from fear appeals in favour of coping instructions and individual incentivisation

For policymakers, moving away from fear appeals in favour of coping instructions and incentivising individuals may be the best way to achieve cybersecurity behavioural change (Chapter 3). These insights were obtained by combining psychological and behavioural economics approaches: Drawing on psychological models such as the Extended Parallel Process Model revealed that fear appeals are not likely to trigger behaviour that will be effective in reducing cybersecurity threats.

Instead, giving behavioural recommendations such as coping instructions is much more likely to be successful. Policymakers should employ interventions that provide individuals with an actionable response to a threat in order to achieve meaningful behavioural change. In addition, Prospect Theory and other behavioural economics models underscore the importance of utilising incentives to drive better cybersecurity behaviour. Policymakers should examine these models to determine how to structure a system of rewards if they are to maximise their potential impact on cybersecurity behaviour.

Aiming policy interventions at individuals' System 1-type thinking

To have a significant impact on an individuals' behaviour, policy interventions should be aimed at individuals' perceptions of cyber attack probabilities (Chapter 3). Drawing on Prospect Theory, we learned that individuals' cybersecurity and cyber insurance decisions are not based on their actual risk (i.e. probability) of experiencing a cyber attack, even if accurately assessing that risk were possible. Instead, they are based on the individuals' interpretations of the probability, which are called "decision weights". (The probabilities can be transformed into decision weights using a weighting function.) Since this is very much based on System 1-type thinking, we can use BEEs in order to assess individuals' decision weights. This suggests that policy interventions should target individuals' decision weights (and/or the weighting function used to obtain those decision weights) in order to affect individuals' decisions regarding cybersecurity behaviour and/or the purchase of cybersecurity products or cyber insurance.

Leveraging the synergy of combined policy options

It is important to determine the best way to leverage the combined effect of policy options, for maximum policy effectiveness. We used an Agent-Based Model (ABM) to examine the effects of different policy interventions on the ecosystem as a whole, combined with a Synergy Experiment to assess the effect of two or more discrete policy interventions acting together to create an effect greater than the sum of its parts (Chapter 2). This revealed that individual policy options had only modest effects on the average cybersecurity level of the ecosystem, with the most observable effects resulting from the Synergy Experiment. Policymakers should therefore investigate how to best combine policy measures for maximum impact.

6.5 Conclusions

We have presented an overview of the key points in this book, highlighted some of its new and innovative approaches to cybersecurity and cyber insurance, and also discussed the policy implications of our research. Throughout, we have focused on how to increase the uptake of cyber insurance. We have also recognised the importance of incentives in this book, and that cyber insurance has the potential to play a key role in encouraging companies to improve their cybersecurity levels, notably in exchange for reduced premiums. In order to have the hoped-for effect, cyber insurance still needs to be adopted on a wider scale. We therefore hope that this volume will play a role in promoting the growth of cyber insurance, thereby helping reduce overall cybersecurity risk for society as a whole.

Bibliography

Accenture and Ponemon Institute (2019). The cost of cyber-crime. Available at: https://www.accenture.com/_acnmedia/PDF-96/Accenture-2019-Cost-of-Cybercrime-Study-Final.pdf#zoom=50.

Adroit Market Research (2019). Global cyber security insurance market 2025.

Advisen (2015). 2015 network security and cyber risk management: The fourth annual survey of enterprise-wide cyber risk management practices in Europe.

Ajzen, I. (1991). The theory of planned behavior. *Organizational Behavior and Human Decision Processes*, 50(2):179–211.

Ajzen, I. and Madden, T. J. (1986). Prediction of goal-directed behavior: Attitudes, intentions, and perceived behavioral control. *Journal of Experimental Social Psychology*, 22(5):453–474.

Allodi, L. and Massacci, F. (2017). Security events and vulnerability data for cybersecurity risk estimation. *Risk Analysis*, 37(8):1606–1627.

Alventosa, A., Gómez, Y., Martínez-Molés, V., and Vila, J. (2016). Location and innovation optimism: A behavioral-experimental approach. *Journal of the Knowledge Economy*, 7(4):890–904.

Amutio, M., Candau, J., and Mañas, J. (2012). *MAGERIT – Versión 3.0. Methodology for Information Systems Risk Analysis and Management*. Available at: https://administracionelectronica.gob.es/pae_Home/pae_Documentacion/pae_Metodolog/pae_Magerit.html?idioma=en.

Anderson, R. and Fuloria, S. (2010). Security economics and critical national infrastructure. In *Economics of Information Security and Privacy*, pages 55–66. Springer.

ANSSI (2010). Expression des Besoins et Identification des Objectifs de Sécurité (EBIOS). Agence Nationale de la Sécurité des Systèmes d'Information (ANSSI). Available at: https://www.ssi.gouv.fr/uploads/2011/10/EBIOS-1-GuideMethodologique-2010-01-25.pdf.

Aon Inpoint (2018). US cyber market update: 2018 US cyber insurance profits and performance. Available at: http://thoughtleadership.aon.com/Documents/201906-us-cyber-market-update.pdf.

Awondo, S. (2019). Efficiency of region-wide catastrophic weather risk pools: Implications for African Risk Capacity insurance program. *Journal of Development Economics*, 136(3):111–118.

Aytes, K. and Connolly, T. (2004). Computer security and risky computing practices: A rational choice perspective. *Journal of Organizational and End User Computing*, 16(3):22–40.

Baer, W. S. and Parkinson, A. (2007). Cyberinsurance in IT security management. *IEEE Security & Privacy*, 5(3):50–56.

Bailey, L. (2014). Mitigating moral hazard in cyber-risk insurance. *Journal of Law & Cyber Warfare*, 3(1):1–42.

Balchanos, M. G. (2012). *A Probabilistic Technique for the Assessment of Complex Dynamic System Resilience*. PhD thesis, Georgia Institute of Technology.

Bandyopadhyay, T., Mookerjee, V. S., and Rao, R. C. (2009). Why IT managers don't go for cyber-insurance products. *Communications of the ACM*, 52(11):68–73.

Banks, D., Ríos, J., and Ríos Insua, D. (2015). *Adversarial Risk Analysis*. Chapman and Hall/CRC Press.

Beautement, A., Becker, I., Parkin, S., Krol, K., and Sasse, A. (2016). Productive security: A scalable methodology for analysing employee security behaviours. In *Twelfth Symposium on Usable Privacy and Security*, pages 253–270.

Bedford, T. and Cooke, R. (2001). *Probabilistic Risk Analysis: Foundations and Methods*. Cambridge University Press.

Berr, J. (2016). WannaCry ransomware attack losses could reach $4 billion. Available at: https://www.cbsnews.com/news/wannacry-ransomware-attacks-wannacry-virus-losses/.

Brenner, J. F. (2013). Eyes wide shut: The growing threat of cyber attacks on industrial control systems. *Bulletin of the Atomic Scientists*, 69(5):15–20.

Bulgurcu, B., Cavusoglu, H., and Benbasat, I. (2010). Information security policy compliance: An empirical study of rationality-based beliefs and information security awareness. *MIS Quarterly*, 34(3):523–548.

Burke, W. W. and Litwin, G. H. (1992). A causal model of organizational performance and change. *Journal of Management*, 18(3):523–545.

Campbell, J., Ma, W., and Kleeman, D. (2011). Impact of restrictive composition policy on user password choices. *Behaviour & Information Technology*, 30(3):379–388.

Cardenas, A., Amin, S., and Sastry, S. (2008). Research challenges for the security of control systems. In *Proceedings of the 3rd USENIX Workshop on Hot Topics in Security*.

Cardenas, A., Amin, S., Sinopoli, B., Giani, A., Perrig, A., and Sastry, S. (2009). Challenges for securing cyber physical systems. In *Workshop on Future Directions in Cyber-Physical Systems Security*.

Carmo Farinha, L. M. (2015). *Handbook of Research on Global Competitive Advantage through Innovation and Entrepreneurship*. IGI Global.

Cavusoglu, H., Raghunathan, S., and Yue, W. T. (2008). Decision-theoretic and game-theoretic approaches to IT security investment. *Journal of Management Information Systems*, 25(2):281–304.

CCTA (2003). CCTA risk analysis and management method (CRAMM). Central Computer and Telecommunications Agency (CCTA).

Chellappa, R. K. and Sin, R. G. (2005). Personalization versus privacy: An empirical examination of the online consumer's dilemma. *Information Technology and Management*, 6(2-3):181–202.

Cialdini, R. B., Demaine, L. J., Sagarin, B. J., Barrett, D. W., Rhoads, K., and Winter, P. L. (2006). Managing social norms for persuasive impact. *Social Influence*, 1(1):3–15.

Cihon, P., Guitierrez, G. M., Kee, S., Kleinaltenkamp, M., and Voigt, T. (2018). Why certify? Increasing adoption of the proposed EU Cybersecurity Certification Framework.

Clemen, R. T. and Reilly, T. (2013). *Making Hard Decisions.* Cengage Learning.

Cloud Security Alliance (2019). Cloud controls matrix (CCM), Version 3.0.1. Available at: `https://cloudsecurityalliance.org/artifacts/cloud-controls-matrix-v3-0-1/`.

Common Criteria (2017). Common criteria for information technology security evaluation (CC), Version 3.1 Release 5. Available at: `https://www.commoncriteriaportal.org/cc/`.

Cooke, R. and Goosens, L. (2000). Procedures guide for structural expert judgement in accident consequence modelling. *Radiation Protection Dosimetry*, 90(3):303–309.

Couce-Vieira, A., Rios Insua, D., and Kosgodagan, A. (2020a). Assessing and forecasting cybersecurity impacts. *Decision Analysis*.

Couce-Vieira, A., Rios Insua, D., Koutalieris, G., and Chatzigiannakis, V. (2020b). A decision support system for cybersecurity risk management. Technical report, ICMAT-CSIC. Available as Appendix 1 at `https://www.cybeco.eu/images/items/CYBECO-D3.2_Improved\%20Modelling\%20framework\%20for\%20cyber\%20risk\%20management_v2.0.pdf`.

Cox, L. A. (2008). What's wrong with risk matrices? *Risk Analysis*, 28(2):497–512.

Cummins, J. D. and Tennyson, S. (1996). Moral hazard in insurance claiming: Evidence from automobile insurance. *Journal of Risk and Uncertainty*, 12(1):29–50.

Davinson, N. and Sillence, E. (2010). It won't happen to me: Promoting secure behaviour among internet users. *Computers in Human Behavior*, 26(6):1739–1747.

de Bruijn, H. and Janssen, M. (2017). Building cybersecurity awareness: The need for evidence-based framing strategies. *Government Information Quarterly*, 34(1):1–7.

Dodel, M. and Mesch, G. (2017). Cyber-victimization preventive behavior: A health belief model approach. *Computers in Human behavior*, 68(1):359–367.

Ducos, B. and de Ligniéres, L. (December 2019). Cyberinsurance, a risk nearly like the others? *Les Annales des Mines*, No. 8. Available at: `http://annales.org/enjeux-numeriques/2019/resumes/decembre/16-en-resum-FR-AN-decembre-2019.html#16AN`.

Earp, J. B. and Baumer, D. (2003). Innovative web use to learn about consumer behavior and online privacy. *Communications of the ACM*, 46(4):81–83.

EIOPA (2018). Understanding cyber insurance – A structured dialogue with insurance companies. European Insurance and Occupational Pensions Authority (EIOPA).

EIOPA (2019). Cyber risk for insurers – Challenges and opportunities. European Insurance and Occupational Pensions Authority (EIOPA).

Ekin, T. (2020). *Statistics and Health Care Fraud.* Chapman and Hall/CRC Press.

Eling, M. and Jung, K. (2018). Copula approaches for modeling cross-sectional dependence of data breach losses. *Insurance: Mathematics and Economics*, 82:167–180.

Eling, M. and Schnell, W. (2016). What do we know about cyber risk and cyber risk insurance? *The Journal of Risk Finance*, 17(5):474–491.

Eling, M. and Wirfs, J. (2019). What are the actual costs of cyber risk events? *European Journal of Operational Research*, 272(3):1109–1119.

ENISA (2019). Cybersecurity culture guidelines: Behavioural aspects of cybersecurity. European Union Agency for Cybersecurity (EN-ISA). Available at: https://www.enisa.europa.eu/publications/cybersecurity-culture-guidelines-behavioural-aspects-of-cybersecurity.

Estamos Seguros (2017). Cada minuto revientan cuatro cañerías en España. Available at: https://www.estamos-seguros.es/cada-minuto-revientan-cuatro-canerias-en-espana/.

European Commission (2020). What is an SME? Available at: http://ec.europa.eu/growth/smes/business-friendly-environment/sme-definition.

Evans, J. S. B. (2003). In two minds: Dual-process accounts of reasoning. *Trends in Cognitive Sciences*, 7(10):454–459.

Evans, M., He, Y., Yevseyeva, I., and Janicke, H. (2018). Analysis of published public sector information security incidents and breaches to establish the proportions of human error. In *Proceedings of the Twelfth International Symposium on Human Aspects of Information Security & Assurance*, pages 191–199. University of Plymouth.

Farquhar, P. (1984). State of the art – Utility assessment methods. *Management Science*, 30(11):1268–1300.

Fauntleroy, J. C., Wagner, R. R., and Odell, L. A. (2015). Cyber insurance – Managing cyber risk. Technical report, Institute for Defense Analyses.

Fielder, A., Panaousis, E., Malacaria, P., Hankin, C., and Smeraldi, F. (2016). Decision support approaches for cyber security investment. *Decision Support Systems*, 86:13–23.

Floyd, D. L., Prentice-Dunn, S., and Rogers, R. W. (2000). A meta-analysis of research on protection motivation theory. *Journal of Applied Social Psychology*, 30(2):407–429.

French, S. and Insua, D. R. (2000). *Statistical Decision Theory.* Wiley.

Furman, S., Theofanos, M. F., Choong, Y.-Y., and Stanton, B. (2011). Basing cybersecurity training on user perceptions. *IEEE Security & Privacy*, 10(2):40–49.

Furnell, S. (2007). An assessment of website password practices. *Computers & Security*, 26(7-8):445–451.

George, J. M. (1992). Extrinsic and intrinsic origins of perceived social loafing in organizations. *The Academy of Management Journal*, 35(1):191–202.

GlobalDots (2019). DDoS distributed denial-of-service explained. Available at: `https://www.globaldots.com/ddos-distributed-denial-of-service-explained`.

Greenberg, A. (2018). The untold story of NotPetya, the most devastating cyberattack in history. *Wired*.

Han, J., Kim, Y. J., and Kim, H. (2017). An integrative model of information security policy compliance with psychological contract: Examining a bilateral perspective. *Computers & Security*, 66(1):52–65.

Hargrave, M. (2019). Middle market firm. Available at: `https://www.investopedia.com/terms/m/middle-market-firms.asp`.

Heidt, M., Gerlach, J., and Buxmann, P. (2019). A holistic view on organizational IT security: The influence of contextual aspects during IT security decisions. In *Proceedings of the 52nd Hawaii International Conference on System Sciences*, pages 6145–6154.

Henson, R. and Garfield, J. (2016). What attitude changes are needed to cause SMEs to take a strategic approach to information security? *Athens Journal of Business and Economics*, 2(3):303–318.

Herath, T. and Rao, H. R. (2009). Protection motivation and deterrence: A framework for security policy compliance in organisations. *European Journal of Information Systems*, 18(2):106–125.

Herley, C. and Florêncio, D. (2010). Nobody sells gold for the price of silver: Dishonesty, uncertainty and the underground economy. In *Economics of Information Security and Privacy*, pages 33–53. Springer.

Hernández, P. and Vila, J. (2014). Measuring value levers: Experimental and contingent approaches. *Journal of Business Research*, 67(5):775–778.

Holt, C. A. and Laury, S. K. (2002). Risk aversion and incentive effects. *American Economic Review*, 92(5):1644–1655.

Hubbard, D. W. and Seiersen, R. (2016). *How to measure anything in cybersecurity risk*. Wiley.

Imperva (2019). 2019 global DDoS threat landscape report. Available at: `https://www.imperva.com/resources/resource-library/reports/global-ddos-threat-landscape`.

Inpoint, A. (2017). Global cyber market overview: Uncovering the hidden opportunities.

ISF (2017). Information risk assessment methodology 2. Information Security Forum (ISF).

ISO (2011). ISO/IEC 27005 – Information security risk management. International Organization for Standardization (ISO).

ISO (2013). ISO/IEC 27001 – Information security management systems – Requirements. International Organization for Standardization (ISO).

Jansen, J. and van Schaik, P. (2019). The design and evaluation of a theory-based intervention to promote security behaviour against phishing. *International Journal of Human-Computer Studies*, 123(1):40–55.

Jeske, D., Briggs, P., and Coventry, L. (2016). Exploring the relationship between impulsivity and decision-making on mobile devices. *Personal and Ubiquitous Computing*, 20(4):545–557.

Kahneman, D. (2011). *Thinking, Fast and Slow*. Macmillan.

Kankanhalli, A., Teo, H.-H., Tan, B. C., and Wei, K.-K. (2003). An integrative study of information systems security effectiveness. *International Journal of Information Management*, 23(2):139–154.

Kesan, J. P. and Hayes, C. M. (2017). Strengthening cybersecurity with cyberinsurance markets and better risk assessment. *Minnesota Law Review*, 102(1):191–275.

Khalili, M. M., Naghizadeh, P., and Liu, M. (2018). Designing cyber insurance policies: The role of pre-screening and security interdependence. *IEEE Transactions on Information Forensics and Security*, 13(9):2226–2239.

Khouzani, M., Liu, Z., and Malacaria, P. (2019). Scalable min-max multi-objective cybersecurity optimisation over probabilistic attack graphs. *European Journal of Operational Research*, 278(3):894–903.

Kirlappos, I., Parkin, S., and Sasse, M. A. (2014). Learning from "shadow security": Why understanding non-compliance provides the basis for effective security. In *Proceedings Workshop on Usable Security*.

Kirlappos, I., Sasse, M. A., and Harvey, N. (2012). Why trust seals don't work: A study of user perceptions and behavior. In *International Conference on Trust and Trustworthy Computing*, pages 308–324. Springer.

Kissel, R. (2013). Glossary of key information security terms. *National Institute of Standards and Technology Interagency Reports (NIST IR)*, 7298(3).

Klahr, R., Shah, J., Sheriffs, P., Rossington, T., Pestell, G., Button, M., and Wang, V. (2017). Cyber security breaches survey 2017. UK Department for Culture, Media, and Sport.

Kuru, D. and Bayraktar, S. (2017). The effect of cyber-risk insurance to social welfare. *Journal of Financial Crime*, 24(2):329–346.

Labunets, K., Pieters, W., van Gelder, P., van Eeten, M., Branley-Bell, D., Briggs, P., Coventry, L., Vila, J., and Gomez, Y. (2019). D7.1: CYBECO policy recommendations. Available at: `https://www.cybeco.eu/results`.

Leach, J. (2003). Improving user security behaviour. *Computers & Security*, 22(8):685–692.

Levi-Faur, D. (2011). *Handbook on the Politics of Regulation*. Edward Elgar Publishing.

Li, H., Luo, X. R., Zhang, J., and Sarathy, R. (2018). Self-control, organizational context, and rational choice in internet abuses at work. *Information & Management*, 55(3):358–367.

Lloyd's and University of Cambridge Centre for Risk Studies (2015). Business blackout: The insurance implications of a cyber attack on the US power grid. Available at: `https://www.jbs.cam.ac.uk/fileadmin/user_upload/research/centres/risk/downloads/crs-lloyds-business-blackout-scenario.pdf`.

Low, P. (2017). Insuring against cyber-attacks. *Computer Fraud & Security*, 2017(4):18–20.

Lund, M. S., Solhaug, B., and Stølen, K. (2011). *Model-Driven Risk Analysis: The CORAS Approach*. Springer-Verlag Berlin Heidelberg.

Manworren, N., Letwat, J., and Daily, O. (2016). Why you should care about the target data breach. *Business Horizons*, 59(3):257–266.

Marinos, L. and Lourenço, M. (2019). Threat landscape report 2018. European Union Agency for Cybersecurity (ENISA). Available at: `https://www.enisa.europa.eu/publications/enisa-threat-landscape-report-2018`.

Marotta, A., Martinelli, F., Nanni, S., Orlando, A., and Yautsiukhin, A. (2017). Cyber-insurance survey. *Computer Science Review*, 24(1):35–61.

Martinez Bustamante, I. (2018). Drivers and impediments for cyber insurance adoption among Dutch SMEs. Master's thesis, Delft University of Technology. Available at: `https://repository.tudelft.nl/islandora/object/uuid:4503fb87-2a46-4ea4-883a-6b48a4a345bc?collection=education`.

McAfee and Center for Strategic and International Studies (2018). Economic impact of cybercrime – No slowing down. Available at: `https://www.mcafee.com/enterprise/en-us/assets/reports/restricted/rp-economic-impact-cybercrime.pdf`.

Meland, P. H., Tondel, I. A., and Solhaug, B. (2015). Mitigating risk with cyberinsurance. *IEEE Security & Privacy*, 13(6):38–43.

Mentzer, J. T., DeWitt, W., Keebler, J. S., Min, S., Nix, N. W., Smith, C. D., and Zacharia, Z. G. (2001). Defining supply chain management. *Journal of Business Logistics*, 22(2):1–25.

Miyazaki, A. D. and Fernandez, A. (2001). Consumer perceptions of privacy and security risks for online shopping. *The Journal of Consumer Affairs*, 35(1):27–44.

National Technical Authority for Information Assurance (2012). HMG information assurance standard no. 1 (IS1).

Naveiro, R., Redondo, A., Rios Insua, D., and Ruggeri, F. (2019). Adversarial classification: An adversarial risk analysis approach. *International Journal of Approximate Reasoning*, 113(2):133–148.

Nexus, C. (2016). State of cybersecurity: Implications for 2016. In *An ISACA and RSA Conference Survey*.

Nicholson, J., Coventry, L., and Briggs, P. (2018). Introducing the cybersurvival task: Assessing and addressing staff beliefs about effective cyber protection. In *Fourteenth Symposium on Usable Privacy and Security*, pages 443–457.

NIST (2012). Special Publication (SP) 800-30, Revision 1. Guide for conducting risk assessments. National Institute of Standards and Technology (NIST). Available at: `https://nvlpubs.nist.gov/nistpubs/Legacy/SP/nistspecialpublication800-30r1.pdf`.

NIST (2018). Special Publication (SP) 800-37, Revision 2. Risk management framework for information systems and organizations: A system life cycle approach for security and privacy. National Institute of Standards and Technology (NIST). Available at: `https://nvlpubs.nist.gov/nistpubs/SpecialPublications/NIST.SP.800-37r2.pdf`.

Pahnila, S., Siponen, M., and Mahmood, A. (2007). Employees' behavior towards IS security policy compliance. In *2007 40th Annual Hawaii International Conference on System Sciences*, pages 156b–156b. IEEE.

Pal, R., Golubchik, L., Psounis, K., and Hui, P. (2017). Security pricing as an enabler of cyber-insurance: A first look at differentiated pricing markets. *IEEE Transactions on Dependable and Secure Computing*, 16(2):358–372.

Pfleeger, S. L. and Caputo, D. D. (2012). Leveraging behavioral science to mitigate cyber security risk. *Computers & Security*, 31(4):597–611.

Positive Technologies (2018). Social engineering: How the human factor puts your company at risk. Available at: `https://www.ptsecurity.com/ww-en/analytics/social-engineering-2018/`.

P&S Intelligence (2017). Cyber insurance market by enterprise size (large enterprise, small and medium enterprise), by service, by industry vertical, by geography – Global market size, share, development, growth and demand forecast, 2013-2023.

Rao, N. S., Poole, S. W., Ma, C. Y., He, F., Zhuang, J., and Yau, D. K. (2016). Defense of cyber infrastructures against cyber-physical attacks using game-theoretic models. *Risk Analysis*, 36(4):694–710.

Reynolds, C. (2019). Misconfigured servers still a key risk for companies moving to the cloud. Available at: `https://www.cbronline.com/news/misconfigured-servers`.

Rios Insua, D., Couce-Vieira, A., Rubio, J. A., Pieters, W., Labunets, K., and Rasines, D. G. (2019). An adversarial risk analysis framework for cybersecurity. *Risk Analysis*.

Robinson, N. (2012). Incentives and barriers of the cyber insurance market in europe. European Union Agency for Cybersecurity (ENISA). Available at: `https://www.enisa.europa.eu/publications/incentives-and-barriers-of-the-cyber-insurance-market-in-europe`.

Rockafellar, T. and Uryasev, S. (2002). Conditional value-at-risk for general loss distributions. *Journal of Banking and Finance*, 26(7):1443–1471.

Romanosky, S., Ablon, L., Kuehn, A., and Jones, T. (2019). Content analysis of cyber insurance policies: How do carriers price cyber risk? *Journal of Cybersecurity*, 5(1):1–19.

Ruiter, R. A., Kessels, L. T., Peters, G.-J. Y., and Kok, G. (2014). Sixty years of fear appeal research: Current state of the evidence. *International Journal of Psychology*, 49(2):63–70.

Sarabia, J. M., Gomez-Deniz, E., and Vazquez-Polo, F. (2007). *Estadística Actuarial: Teoria y Aplicaciones*. Pearson Education.

Sasse, M. A. and Flechais, I. (2005). Usable security: Why do we need it? How do we get it? In *Security and Usability: Designing Secure Systems that People Can Use*, pages 13–30. O'Reilly.

Schilling, A. and Werners, B. (2016). Optimal selection of IT security safeguards from an existing knowledge base. *European Journal of Operational Research*, 248(1):318–327.

Sewnandan, J. (2018). Analysing the impact of cyber insurance on the cyber security ecosystem: Utilising agent-based modelling to explore the effects of insurance policies. Master's thesis, Delft University of Technology. Available at: `https://repository.tudelft.nl/islandora/object/uuid\%3Acabcc9b3-9814-4c05-940e-6916baa9adf8?collection=education`.

Sharma, Y. (2018). Cyber insurance market to reach \$14 billion, globally, by 2022. Allied Market Research.

Sklar, A. (1959). Fonctions de répartition à n dimensions et leurs marges. *Publications de l'Institut Statistique de l'Université de Paris*, 8(3):229–231.

Smith, V. L. and Smith, V. (1991). *Papers in Experimental Economics*. Cambridge University Press.

Stafford, T. (2017). On cybersecurity loafing and cybercomplacency. *ACM SIGMIS Database: The Database for Advances in Information Systems*, 48(3):8–10.

Tech!Espresso (2020). Liquid damaged laptop or device? Available at: `https://techespresso.ca/computer-liquid-spill-repair-calgary-0.html`.

Thomas, P. (2013). The risk of using risk matrices. Master's thesis, University of Stavanger.

Tisue, S. and Wilensky, U. (2004). Netlogo: A simple environment for modeling complexity. In *International Conference on Complex Systems*, volume 21, pages 16–21.

Tolvanen, J. (2015). Measuring moral hazard using insurance panel data. Available at: `http://www.eief.it/files/2016/01/02-jmp_tolvanen.pdf`.

Torres, A., Redondo, A., Rios Insua, D., Domingo, J., and Ruggeri, F. (2020). Structured expert judgement in supply chain cyber risk management. In *Expert Judgement in Risk and Decision Analysis (Hanea, Nane, Bedford, French, eds)*. Springer.

Turland, J., Coventry, L., Jeske, D., Briggs, P., and van Moorsel, A. (2015). Nudging towards security: Developing an application for wireless network selection for Android phones. In *Proceedings of the 2015 British HCI Conference*, pages 193–201. Association for Computing Machinery (ACM).

Tversky, A. and Kahneman, D. (1992). Advances in prospect theory: Cumulative representation of uncertainty. *Journal of Risk and Uncertainty*, 5(4):297–323.

Valls, F. (2019). Multa de €60,000 a RTVE por el robo de 6 USBs con datos bancarios de la plantilla. Available at: `https://www.lainformacion.com/empresas/rtve-multa-usb-bancos-trabajadores-prejubilados/6524036/`.

van Bavel, R., Rodríguez-Priego, N., Vila, J., and Briggs, P. (2019). Using protection motivation theory in the design of nudges to improve online security behavior. *International Journal of Human-Computer Studies*, 123(1):29–39.

van Dam, K. H., Nikolic, I., and Lukszo, Z., editors (2013). *Agent-based modelling of socio-technical systems*, volume 9 of *Agent-Based Social Systems*. Springer.

Verisign (2017). Q1 2017 DDoS trends report: 26 percent increase in average peak attack size. Available at: `https://blog.verisign.com/security/q1-2017-ddos-trends-report-26-percent-increase-in-average-peak-attack-size/`.

Verma, R. and Marchette, D. (2019). *Cybersecurity Analytics*. Chapman and Hall/CRC Press.

Warkentin, M., Johnston, A. C., Walden, E., and Straub, D. W. (2016). Neural correlates of protection motivation for secure IT behaviors: An fMRI examination. *Journal of the Association for Information Systems*, 17(3):194–215.

WEF (2020). The global risks report. World Economic Forum (WEF). Available at: `https://www.weforum.org/reports/the-global-risks-report-2020`.

Weinstein, N. D. (1980). Unrealistic optimism about future life events. *Journal of Personality and Social Psychology*, 39(5):806–820.

Weishäupl, E., Yasasin, E., and Schryen, G. (2018). Information security investments: An exploratory multiple case study on decision-making, evaluation and learning. *Computers & Security*, 77:807–823.

White, T. B. (2004). Consumer disclosure and disclosure avoidance: A motivational framework. *Journal of Consumer Psychology*, 14(1-2):41–51.

Witte, K. (1996). Fear as motivator, fear as inhibitor: Using the extended parallel process model to explain fear appeal successes and failures. In *Handbook of Communication and Emotion*, pages 423–450. Elsiever.

Woods, D. and Simpson, A. (2017). Policy measures and cyber insurance: A framework. *Journal of Cyber Policy*, 2(2):209–226.

Young, D., Lopez Jr., J., Rice, M., Ramsey, B., and McTasney, R. (2016). A framework for incorporating insurance in critical infrastructure cyber risk strategies. *International Journal of Critical Infrastructure Protection*, 14(1):43–57.

Zhang, X. A. and Borden, J. (2019). How to communicate cyber-risk? An examination of behavioral recommendations in cybersecurity crises. *Journal of Risk Research*.

Index

Printed in the United States
By Bookmasters